Simulation Based Engineering in Solid Mechanics

J.S. Rao

Simulation Based Engineering in Solid Mechanics

Springer

J.S. Rao
The Vibration Institute of India
Bangalore
India

ISBN 978-3-319-47613-1 ISBN 978-3-319-47614-8 (eBook)
DOI 10.1007/978-3-319-47614-8

Library of Congress Control Number: 2016955327

Printed on acid-free paper

This Springer imprint is published by Springer Nature
The registered company is Springer International Publishing AG
The registered company address is: Gewerbestrasse 11, 6330 Cham, Switzerland

To my beloved wife Indira
and to the new arrival in the family great
grandson Rumi
and in memory of my parents
Jammi Chikka Rao and Jammi Ramanamma
and in memory of my mentor
Dr. Nachimuthu Mahalingam

Preface

Solid mechanics is one of the oldest subjects practiced since antique time.

We built monuments without the knowledge of structures as we know them today. To the best we know, beams and columns were used in building places of worship for five or more millennia. We probably have guessed that a column or a pillar takes an axial load corresponding to an equally apportioned weight of the structure along its axis and used our intuition to provide a cut rock across the top of the pillars, which we began calling a beam. Though we did not know how a beam works, we apparently developed rules based on the experience. This curiosity made man think on how a beam works without knowledge of deformation, strain, or stress, yet we have probably seen the structures failing by improper designs or by forces which are not of regular occurrence such as an earthquake or tidal waves.

It is left to the famous Italian thinker Leonardo da Vinci of the medieval period to observe that a beam bends under a transverse load with a plane cross section remaining plane after the load application but rotating about a plane normal to the load application and axis of the beam. This is a profound discovery and remained almost a mystery until Newton (simultaneously Leibniz) came up with the discovery of Calculus that ushered in the Scientific Revolution.

Lagrange embedded the universal energy conservation principle and opened another branch of what we call today Lagrangian Mechanics as against Newtonian Mechanics. Euler and Lagrange applied this energy principle to beams (other structures' definition is not known at that time) and provided an alternate to Newtonian Mechanics in the form of Lagrangian Mechanics which remains even today the most celebrated principle in different kinds of physics problems including that of structures. This subject is called *elasticity* and is quite complicated with a solution to be obtained by solving 15 coupled partial differential equations as long as the structure is not experiencing permanent deformations or *plasticity*; then, the problem becomes somewhat more complicated.

During the Science Revolution period, basic governing equations for solids were derived; however, they remained unsolvable with the numerical tools available at that time. Moreover, the need for solutions of these equations was not warranted because even during the industrial revolution the slow-speed reciprocating steam engines could be designed based on the trial-and-error tests.

During the eighteenth and nineteenth centuries, there were practically no engineering applications except for James Watt ushering in the industrial revolution with a reciprocating steam engine. The reciprocating steam engine being very slow in operation did not warrant stress and flow design of a high level.

Strangely, it was the desire to understand how drums beat and produce sound that led Sophie Germain to provide the first theory of Plates that was followed by Theory of Shells, which are still simplifications of actual structures that were studied mainly by physicists.

This scenario changed with the construction of a practical dynamo by Thomas Alva Edison and construction of rotating machinery, first an impulse turbine by De Laval and reaction turbine by Charles Parsons toward the end of nineteenth century. (Actually Hero of Alexandria built nearly 2000 years ago the first reaction turbine but had no idea what to do with it beyond opening the temple doors using pulleys invented by Archimedes in that period). The initial sizes of these turbines were so small in size, producing few KW of power, that they could be built by an intuition and trial-and-error approach.

Man has quickly learnt to adapt to the luxury of electricity and began demanding large-size machines that can produce more power; they in turn demanded designs which needed a better understanding of stress (and strain). To achieve these designs, there were no numerical tools to handle the complex physics of solids.

It was the genius of Rayleigh toward the end of the nineteenth century and Stodola at the beginning of the twentieth century who developed approximate methods to solve the elasticity problems, which became to be known as Strength of Materials. It is these approximate methods that stood the test of time for over a century in determining the average stress and apply stress concentration factors to determine the peak stress. The peak stress thus determined was coupled with a factor of safety (or some prefer to call it a factor of ignorance) and accomplish the designs, and the prototypes were built and tested.

Professor Stephen Timoshenko wrote the first version of his famous Strength of Materials textbook during 1907 to 1911 at Kyiv Polytechnic Institute in Ukraine soon after the work of Rayleigh. He moved to USA in 1922 and joined Westinghouse, an exciting place to be during the twenties. This was a period of industrial growth and expansion providing challenges in mechanics where Timoshenko became the most influential person. The five years that Timoshenko spent at Westinghouse have been called its "Golden Era of Mechanics." His book on Strength of Materials was written in English in two volumes and published in 1930. Subsequently, he abridged these two volumes into one and published Elements of Strength of Materials, primarily addressing the undergraduate students in American colleges in 1935. Since then, many books have appeared on this subject matter.

The invention of the digital computer using valves in Philadelphia (ENIAC) during the Second World War changed the scenario gradually; this was accelerated by transistors and subsequently integrated circuits in 1960s and it was all of a sudden a boom for number crunching. The high-performance computing that followed gradually replaced engineering approximations through Strength of Materials by Elasticity and Lagrangian approach. It is now "Science to Engineering" or "Simulation-Based Engineering Science" through several commercially available codes to do the drudgery arising out of finite element methods or finite volume methods. This revolutionized the outlook for designs; what used to be months of design time by several skilled engineers is now seconds or minutes of numerical effort on high-speed computers.

There are several books dealing with the subject of finite element methods emanating from Lagrangian Mechanics or Energy Methods for all different kinds of structures—then why this book?

Industry practices have changed drastically in the recent decade or two by adapting the computer-aided methods leaving the approximate methods for design analysis practiced during the twentieth century. The structure was first CAD modeled followed by meshing before applying boundary conditions and setting up numerical problems to be solved by a solver. These aspects of modeling, meshing, and applying the boundary conditions belong to preprocessing stage before a problem is sent to a solver. The results from the solver go to the postprocessor to get what we may call a nearly complete exact picture of stress that goes into design. Most of the educational institutions across the world still follow basic courses in solid mechanics developed during the last century. There is a need to bring science upfront in our undergraduate teaching that leads to engineering analysis in a manner that is directly applicable to industry practices. The methods developed during precomputer era are now becoming history and can be offered as electives at a later stage for those who want to learn historical aspects. We can now use directly Simulation-Based Engineering Science (SBES) with high-performance computing (HPC) background.

The design houses across the world practice cost-cutting methods in preprocessing aspects and even analysis by skilled workers rather than actual designers. A lot of preprocessing work can be even done outside the design companies, and to save costs, these types of jobs are outsourced to developing countries where engineering colleges mushroomed to cater to this category of jobs. Their standards are to be improved by introducing SBES approach so that they become globally employable.

We begin with an introduction and brief historical perspective of structures. We then have a recap of the required mathematical principles on matrices, linear algebra, vectors, and tensors. The science of deformation, strain, and stress of general three-dimensional solids is first presented defining the solid properties at a point under the application of external traction loads. Most common tension test is described, and the solid properties under deformation are presented.

Lagrangian Mechanics is then introduced together with energy principles. Approximate methods (strength of materials) from equilibrium diagrams are also presented. Only one-dimensional structures classified as bars (axial loads), rods (twisting loads), and beams (bending loads) are considered in this book.

We next consider general stress and strain in a body subjected to loads and derive principal stresses and strains and von Mises stress and strain that are used in the design of structures. A suitable deformation field is assumed, and then, strain and stress fields are determined followed by strain energy. Lagrangian solution was used to derive the governing differential equations consistent with assumed deformation field, and solution for deformations, strains, and stresses was obtained.

In each case of the loading, a finite element method is derived and the corresponding stiffness matrix for the element is derived. Assembling procedure of these matrices is then described to obtain the system matrices. For fewer elements, closed form solutions are illustrated.

The advantage of the finite element method was of course stated to be the adoption of a large number of elements to describe the structure as accurately as possible to the actual case. The procedure of modeling, meshing, and preparing the boundary conditions, material properties, and loads—generally called preprocessing—is not described here. The teacher can give practice to students on a suitable platform of the school's choice. From this preprocessed information, the finite elements, and assembly, the entire data are sent to the solver. The solver is essentially a large-size non-homogeneous set of linear equations solver whose output will be deformation field in all the nodes of the finite elements. This numerical data are used to determine the strains and stresses as described, and the results are displayed in a chosen form—called post-processing. A specific case is presented to illustrate this process and the results compared with the Strength of Materials approach.

Coimbatore
March 2016

J.S. Rao

Acknowledgments

Professor Henry Langhaar of Illinois University in early 1960s taught me Energy Methods which laid down my foundations for solid mechanics knowledge base which still remains my corner stone in the work of turbomachinery blading. I was highly fortunate to have been associated with Prof. William Carnegie of University of Surrey in the late 1960s which gave me a real insight into turbine blade dynamics. I am deeply indebted to them for this knowledge base.

It is my large body of dedicated students who worked hard in filling the puzzle I had on integrated approach to lifting of machine components. They used to work in laboratories, computer centers, and my home, and they are more like my family. I am extremely indebted to their devotion to the subject.

I began to realize slowly and steadily that solids, fluids, and electromagnetism are interdisciplinary, and it is astounding to see our forefathers, Newton, Euler, and Faraday, among others, have bequeathed us a rich legacy of science over three centuries ago. In the absence of computational means, engineering disciplines and practices grew separately over the last century; this scenario is rapidly changing in the last decade or two with basic sciences providing accurate solutions, thus giving rise to Simulation-Based Engineering Science to the forefront. This necessitates a rethink on the approach to inculcate approximate methods to engineering students and replace those by SBES.

My teacher and mentor Prof. B.M. Belgaumkar talked to me about this science 54 years ago; frankly I didn't follow him then, and now I am beginning to appreciate his advice. After a continuous exposure to industrial problems in the last two decades, I recall his industrial service after the retirement in the 1970s. I am very grateful for his guidance in my life.

I am thankful to my colleagues Saravana Kumar and Bigil Kumar for solving some problems on the computer. I am thankful to my Kumaraguru College of Technology office and their help in preparing the materials.

Finally, I would like to express deepest gratitude to Late Arutselvar Padmabhushan Dr. Nachimuthu Mahalingam. He is a freedom fighter, Gandhian, educationalist, a great Tamil scholar, and above all a visionary providing guidance to people like me. He shaped my spiritual part of life in the last decade, particularly during last two years. I have the honor to be associated with him in my academics. I thank him very much for giving me the opportunity in promoting scientific thinking and temper to younger generations of faculty and students.

Contents

Chapter 1
Introduction to Solid Mechanics

Our ancestors had rich traditions in religion and philosophy. Religion and belief in God led them to build huge temples to honor them.

The temple of Zeus in Fig. 1.1 was built between 466 and 456 B.C., more than 2500 years ago during the height of Classical Greek architecture and artistic endeavors. It was designed by Libon, an architect from neighboring Elis. The temple was constructed of local shell stone, the predominant architectural style of the time, and the same style as the *Parthenon* in Athens. The temple itself acted merely as a protective home for the real showpiece, the cult statue of Zeus himself. The Olympic Columns left behind in the ruins illustrate the columns that are used; they carry axial loads from the structure transferring them to the ground as shown in Fig. 1.2.

Mayans built their temples in South America, see Fig. 1.3, between 300 and 600 AD. Somnath temple, Fig. 1.4, in Gujarat was built around 1500 years ago. To Prohm a Buddhist temple, Fig. 1.5 was constructed from 1186 AD. A rare inscription there provides statistics on the temple's workers around 80,000, including 2700 officials.

The load bearing elements or structures common to all constructions of buildings then (and now) are beams and columns. It is not known whether there was any understanding on the principles, although it is expected that several rules must have been developed in building them. These structures lasted for long time before natural disasters such as earth quakes or manmade wars destroyed them.

Leonardo Da Vinci (1452–1519), (Fig 1.6), is attributed to be the first to make the crucial observations on beam bending. We can see the familiar argument that plane cross-sections remain plane before and after bending. Da Vinci lacked Hooke's law and calculus to postulate the beam theory as we still practice in structural engineering. His contributions appear in Codex Madrid I, one of two remarkable notebooks that were discovered in 1967 in the National Library of Spain (Madrid), after being misplaced for nearly 500 years.

Nearly a century later, Galileo Galilei (1564–1642), (Fig. 1.7), is accredited to have made further studies on beams, he made some incorrect assumptions in the

J.S. Rao, *Simulation Based Engineering in Solid Mechanics*,
DOI 10.1007/978-3-319-47614-8_1

Fig. 1.1 Temple of Zeus http://www.olympia-greece.org/templeofzeus1.jpg

Fig. 1.2 Olympian Columns https://upload.wikimedia.org/wikipedia/commons/b/b0/Attica_06-13_Athens_25_Olympian_Zeus_Temple.jpg

Fig. 1.3 Mayan Temple http://resources.touropia.com/gfx/d/step-pyramids-of-the-world/chichen_itza.jpg

Fig. 1.4 Somnath Temple http://happymahashivaratri.in/wp-content/uploads/2015/01/somnath-temple-xzoom.jpg

Fig. 1.5 *To Prohm* Temple, Cambodia http://static.flickr.com/36/105298856_de424ba971.jpg?v=0

Fig. 1.6 Leonardo Da Vinci (1452–1519) and Bending of a Beam http://link.springer.com/chapter/10.1007/978-94-007-1165-5_5#page-1

absence of Hooke's Law—Elasticity (1660) which was subsequently provided by Robert Hooke (1635–1703). The scientific revolution is now around the corner with Sir Isaac Newton (1642–1727) that spurred the physics of solids and fluids and heat transfer, the core of engineering principles. Euler in 1750, (Fig. 1.8), nearly 250 years after Da Vinci gave the simple (engineering or technical) beam theory that we practice even today after another 250 years or more; Bridges and buildings designed by precedent until the late 19th century, when the Eiffel Tower and Ferris Wheel demonstrated the validity of the theory on a large scale.

It is just about 2½ centuries that we have been practicing structural mechanics. Buildings and bridges and even some of the most complex elasticity problems from rotating machinery use Bernoulli-Euler beam theory. We will discuss these methods from basic principles to give us a good understanding on how these practices have evolved in modern high performance computer era.

Fig. 1.7 Galileo Galilei (1564–1642)

Fig. 1.8 Leonhard Euler (1707–1783)

For a historical perspective in this book, one may refer to "History of Rotating Machinery Dynamics" a Springer Publication in 2011 by this author.

Before we begin the study of solid mechanics, some basic mathematical and physics principles will be introduced.

1.1 Matrices

Solid mechanics deals with deformations, strains and stresses of bodies within the continuum due to external tractions or loads. In modern methods, we employ finite element methods and determine these quantities in various elements that can be made to constitute the body. The number of discrete elements required may be even millions depending on the complexity; fortunately modern high speed computers allow us to accomplish this job very quickly. This is achieved through setting up a linear system of equations as we will see later in the book. Dealing with a given system of linear equations requires knowledge of matrices and we will begin with an introduction of this subject.

The origins of mathematical matrices lie with the study of systems of simultaneous linear equations. As early as 300 BC Chinese scholars are known to use matrix methods and solve simultaneous equations. Carl Friedrich Gauss (1777–1855) is attributed with the modern matrix method of solution known as Gaussian elimination. The term *matrix* was introduced in 1850 by James Joseph Sylvester.

A rectangular array of numbers of the form $\begin{bmatrix} a_{11} & a_{12} & \cdots & a_{1n} \\ a_{21} & a_{22} & & a_{2n} \\ \cdots & & & \\ a_{m1} & a_{m2} & \cdots & a_{mn} \end{bmatrix}$ is called an $m \times n$ matrix, with m rows and n columns. The rows are counted from the top and columns from the left. Hence $\begin{bmatrix} a_{i1} & a_{i2} & \cdots & a_{in} \end{bmatrix}$ and $\begin{Bmatrix} a_{1j} \\ a_{2j} \\ \vdots \\ a_{mj} \end{Bmatrix}$ represent respectively the ith row and the jth column of the matrix and a_{ij} represents the entry in the ith row and the jth column. Note that the full matrices or row matrices are usually put in rectangular brackets and the column matrices in curved brackets.

Matrices are additive (or subtractive) only when they are of same size $m \times n$. For example

$$[A] = \begin{bmatrix} 2 & 3 & 1 & -1 \\ 4 & 2 & 0 & 5 \\ 2 & 5 & 1 & 4 \end{bmatrix} \quad [B] = \begin{bmatrix} 1 & 2 & -1 & 1 \\ 3 & 4 & 2 & -3 \\ 1 & 3 & 1 & 2 \end{bmatrix}$$

$$\text{then } [A+B] = \begin{bmatrix} 3 & 5 & 0 & 0 \\ 7 & 6 & 2 & 2 \\ 3 & 8 & 2 & 6 \end{bmatrix}$$

Matrix multiplication by a scalar c is simply given as follows:

$$\text{If } [A] = \begin{bmatrix} a_{11} & a_{12} & \cdots & a_{1n} \\ a_{21} & a_{22} & & a_{2n} \\ \cdots & & & \\ a_{m1} & a_{m2} & \cdots & a_{mn} \end{bmatrix} \text{ then } c[A] = \begin{bmatrix} ca_{11} & ca_{12} & \cdots & ca_{1n} \\ ca_{21} & ca_{22} & & ca_{2n} \\ \cdots & & & \\ ca_{m1} & ca_{m2} & \cdots & ca_{mn} \end{bmatrix}.$$

To understand matrix multiplication, consider a set of linear non-homogeneous algebraic equations given by

$$\begin{aligned} &a_{11}x_1 + a_{12}x_2 + \cdots + a_{1n}x_n = b_1 \\ &a_{21}x_1 + a_{22}x_2 + \cdots + a_{2n}x_n = b_2 \\ &\cdots \\ &a_{m1}x_1 + a_{m2}x_2 + \cdots + a_{mn}x_n = b_m \end{aligned}$$

Let $[A] = \begin{bmatrix} a_{11} & a_{12} & \cdots & a_{1n} \\ a_{21} & a_{22} & & a_{2n} \\ \cdots & & & \\ a_{m1} & a_{m2} & \cdots & a_{mn} \end{bmatrix}$, $\{X\} = \begin{Bmatrix} x_1 \\ x_2 \\ \vdots \\ x_n \end{Bmatrix}$ and $\{B\} = \begin{Bmatrix} b_1 \\ b_2 \\ \vdots \\ b_m \end{Bmatrix}$, we can then express the above set of linear algebraic equations in matrix form $[A]\{X\} = \{B\}$. $[A]$ is the matrix of the coefficients, $\{X\}$ are the unknowns to be obtained by solving the linear equations and $\{B\}$ are the known scalar constants in the equations. The product of matrices $[A]$ and $\{X\}$ is matrix $\{B\}$.

In general if $[A] = \begin{bmatrix} a_{11} & a_{12} & \cdots & a_{1n} \\ a_{21} & a_{22} & & a_{2n} \\ \cdots & & & \\ a_{m1} & a_{m2} & \cdots & a_{mn} \end{bmatrix}$ and $[B] = \begin{bmatrix} b_{11} & b_{12} & \cdots & b_{1p} \\ b_{21} & b_{22} & & b_{2p} \\ \cdots & & & \\ b_{n1} & b_{n2} & \cdots & b_{np} \end{bmatrix}$ are two matrices of size $m \times n$ and $n \times p$ respectively the product $[A]$ $[B]$ is given by a matrix $[Q]$ of size $m \times p$. Note that the number of columns of the first matrix is equal to the number of rows of the second matrix.

$$[Q] = [A][B] = \begin{bmatrix} q_{11} & q_{12} & \cdots & q_{1p} \\ q_{21} & q_{22} & & q_{2p} \\ \cdots & & & \\ q_{m1} & q_{m2} & \cdots & q_{mp} \end{bmatrix}$$

In the above equation for every $i = 1, 2 \ldots m$ and $j = 1, 2 \ldots p$, we have

$$q_{ij} = \sum_{k=1}^{n} a_{ik}b_{kj} = a_{i1}b_{1j} + a_{i2}b_{2j} + \cdots + a_{in}b_{nj}$$

Consider the matrices

$$[A] = \begin{bmatrix} 1 & 2 & 0 & 2 \\ 3 & 2 & 1 & 0 \\ 2 & 0 & 2 & 1 \end{bmatrix}, \; [B] = \begin{bmatrix} 1 & 2 \\ 2 & 3 \\ 2 & 4 \\ 3 & 0 \end{bmatrix} \text{ and determine } [A][B] = \begin{bmatrix} q_{11} & q_{12} \\ q_{21} & q_{22} \\ q_{31} & q_{32} \end{bmatrix}$$

The first row of $[A]$ and the first column of $[B]$ are needed to obtain q_{11} and therefore we form $\begin{bmatrix} 1 & 2 & 0 & 2 \\ \times & \times & \times & \times \\ \times & \times & \times & \times \end{bmatrix} \begin{bmatrix} 1 & \times \\ 2 & \times \\ 2 & \times \\ 3 & \times \end{bmatrix} = \begin{bmatrix} q_{11} & \times \\ \times & \times \\ \times & \times \end{bmatrix}$ and obtain $q_{11} = 1 \times 1 + 2 \times 2 + 0 \times 2 + 2 \times 3 = 11$. In a similar manner we obtain the other entries

$$\begin{aligned} q_{21} &= 3 \times 1 + 2 \times 2 + 1 \times 2 + 0 \times 3 = 9 \\ q_{31} &= 2 \times 1 + 0 \times 2 + 2 \times 2 + 1 \times 3 = 9 \\ q_{12} &= 1 \times 2 + 2 \times 3 + 0 \times 4 + 2 \times 0 = 8 \\ q_{22} &= 3 \times 2 + 2 \times 3 + 1 \times 4 + 0 \times 0 = 16 \\ q_{32} &= 2 \times 2 + 0 \times 3 + 2 \times 4 + 1 \times 0 = 12 \end{aligned} \qquad \therefore [A][B] = \begin{bmatrix} 11 & 8 \\ 9 & 16 \\ 9 & 12 \end{bmatrix}$$

Note that $[B]$ is a 4×2 matrix and $[A]$ is a 3×4 matrix, so that we do not have a definition for the product $[B]\,[A]$.

Determinants:

Determinants occur throughout mathematics. In solving linear algebraic equations, determinants of square matrices play an important role. The determinant provides important information about the matrix of coefficients. The system has a unique solution exactly only when the determinant is nonzero; when the determinant is zero there are either no solutions or many solutions. They are also used in solving Eigenvalue problems in linear algebra.

First we define the *minor* M_{ij} of element a_{ij} as the determinant of the sub-matrix obtained by deleting *i*th row and *j*th column. The *co-factor* is a signed minor, $A_{ij} = (-1)^{i+j} M_{ij}$.

The *determinant* of a square matrix $[A]$ is denoted $|A|$. The determinant of the square matrix $\begin{bmatrix} a_{11} & a_{12} & a_{13} \\ a_{21} & a_{22} & a_{23} \\ a_{31} & a_{32} & a_{33} \end{bmatrix}$ is usually denoted $\Delta = \begin{vmatrix} a_{11} & a_{12} & a_{13} \\ a_{21} & a_{22} & a_{23} \\ a_{31} & a_{32} & a_{33} \end{vmatrix}$. It is evaluated by summing up all co-factors of any of the rows or columns. Thus, using the first row of 3×3 square matrix, its determinant is obtained as follows.

$$\Delta = \begin{vmatrix} a_{11} & a_{12} & a_{13} \\ a_{21} & a_{22} & a_{23} \\ a_{31} & a_{32} & a_{33} \end{vmatrix} = a_{11}\begin{vmatrix} a_{22} & a_{23} \\ a_{32} & a_{33} \end{vmatrix} - a_{12}\begin{vmatrix} a_{21} & a_{23} \\ a_{31} & a_{33} \end{vmatrix} + a_{13}\begin{vmatrix} a_{21} & a_{22} \\ a_{31} & a_{32} \end{vmatrix}$$
$$= a_{11}(a_{22}a_{33} - a_{32}a_{23}) - a_{12}(a_{21}a_{33} - a_{31}a_{23}) + a_{13}(a_{21}a_{32} - a_{31}a_{22})$$
$$= (a_{11}a_{22}a_{33} + a_{12}a_{31}a_{23} + a_{13}a_{21}a_{32}) - (a_{11}a_{32}a_{23} + a_{12}a_{21}a_{33} + a_{13}a_{31}a_{22})$$

The evaluation of the determinant only involves addition, subtraction and multiplication. Thus for instance the determinant of a matrix with integer coefficients will be an integer.

Matrix Inversion:

The inverse of any non-zero number k is $\frac{1}{k}$ or k^{-1}. The inverse has a property that $k \times k^{-1} = k^{-1} \times k = 1$. In a similar manner, we can define the inverse of a square matrix. An $n \times n$ square matrix $[A]$ is invertible (also nonsingular or nondegenerate) if there exists another $n \times n$ square matrix $[B]$ such that $[A]\ [B] = [B]\ [A] = [I]$, where $[I]$ denotes the $n \times n$ identity matrix. If this is the case, then $[B]$ is uniquely determined by $[A]$ and is called the *inverse* of $[A]$, denoted by $[A]^{-1}$. A square matrix is singular if its determinant is 0, it is not invertible. Matrix inversion is the process of finding the matrix $[B]$ for a given invertible matrix $[A]$.

Consider two matrices $[A] = \begin{bmatrix} -1 & 1 \\ -2 & 0 \end{bmatrix}$ and $[B] = \begin{bmatrix} 0 & -\frac{1}{2} \\ 1 & -\frac{1}{2} \end{bmatrix}$.

$$[A][B] = \begin{bmatrix} -1 & 1 \\ -2 & 0 \end{bmatrix}\begin{bmatrix} 0 & -\frac{1}{2} \\ 1 & -\frac{1}{2} \end{bmatrix} = \begin{bmatrix} 1 & 0 \\ 0 & 1 \end{bmatrix}$$

Also

$$[B][A] = \begin{bmatrix} 0 & -\frac{1}{2} \\ 1 & -\frac{1}{2} \end{bmatrix}\begin{bmatrix} -1 & 1 \\ -2 & 0 \end{bmatrix} = \begin{bmatrix} 1 & 0 \\ 0 & 1 \end{bmatrix}$$

We find, $[A][B] = [B][A] = [I]$, therefore $[B]$ is the inverse of $[A]$ or $[B] = [A]^{-1}$. Note that all square matrices do not have an inverse.

Let us try to find the inverse of $[A] = \begin{bmatrix} 2 & 0 \\ 4 & 0 \end{bmatrix}$. Let $[B] = \begin{bmatrix} a & b \\ c & d \end{bmatrix}$ is possible inverse. We can use the above rule that $[A][B] = [B][A] = [I]$, and try to find $[B]$.

$[A][B] = \begin{bmatrix} 2 & 0 \\ 4 & 0 \end{bmatrix}\begin{bmatrix} a & b \\ c & d \end{bmatrix} = \begin{bmatrix} 1 & 0 \\ 0 & 1 \end{bmatrix}$ gives $2a = 1$, $2b = 0$, $4a = 0$ and $4b = 1$.

Clearly the results are contradictory and therefore no inverse exists for $[A]$. This is substantiated by the fact that $|A| = 0$ and being a singular matrix, there is no $[A]^{-1}$.

Gauss Elimination Method:

This method is illustrated for a 3 × 3 matrix. For larger size matrices, the method is tedious and that's where computers come to our rescue.

This method involves operating on the rows of a given matrix in order to reduce it to a unit matrix. The operations performed are

- Interchanging two rows
- Multiplying a row by a constant factor, could be negative or fractional, or both
- Adding a multiple of one row to another.

Consider matrix $[A]^{-1} = \begin{bmatrix} 1 & 3 & 3 \\ 1 & 4 & 3 \\ 2 & 7 & 7 \end{bmatrix}$. To determine the inverse $[A]^{-1}$, begin placing the given $[A]$ and unity matrix $[I]$ adjacent to each other $\begin{bmatrix} 1 & 3 & 3 \\ 1 & 4 & 3 \\ 2 & 7 & 7 \end{bmatrix} \begin{bmatrix} 1 & 0 & 0 \\ 0 & 1 & 0 \\ 0 & 0 & 1 \end{bmatrix}$.

Now proceed by changing the columns of $[A]$ left to right to reduce $[A]$ to an upper triangular form $\begin{bmatrix} 1 & p & q \\ 0 & 1 & r \\ 0 & 0 & 1 \end{bmatrix}$ where p, q and r can be any numbers and entries below the leading diagonal must be zero. Identify the rows in both the matrices as $R1$, $R2$ and $R3$ and perform simultaneously the operations shown to achieve the first column in $[A]$ to the required form in the proposed upper triangular form.

$$\begin{bmatrix} 1 & 3 & 3 \\ 1 & 4 & 3 \\ 2 & 7 & 7 \end{bmatrix} \begin{bmatrix} 1 & 0 & 0 \\ 0 & 1 & 0 \\ 0 & 0 & 1 \end{bmatrix} \begin{matrix} \\ R2 - R1 \\ R3 - 2R1 \end{matrix} \Rightarrow \begin{bmatrix} 1 & 3 & 3 \\ 0 & 1 & 0 \\ 0 & 1 & 1 \end{bmatrix} \begin{bmatrix} 1 & 0 & 0 \\ -1 & 1 & 0 \\ -2 & 0 & 1 \end{bmatrix}$$

The next operation below brings the given matrix $[A]$ to upper triangular form

$$\begin{bmatrix} 1 & 3 & 3 \\ 0 & 1 & 0 \\ 0 & 1 & 1 \end{bmatrix} \begin{bmatrix} 1 & 0 & 0 \\ -1 & 1 & 0 \\ -2 & 0 & 1 \end{bmatrix} \begin{matrix} \\ \\ R3 - R2 \end{matrix} \Rightarrow \begin{bmatrix} 1 & 3 & 3 \\ 0 & 1 & 0 \\ 0 & 0 & 1 \end{bmatrix} \begin{bmatrix} 1 & 0 & 0 \\ -1 & 1 & 0 \\ -1 & -1 & 1 \end{bmatrix}$$

Next, we will make the terms above the diagonal to zero by the following operations.

$$\begin{bmatrix} 1 & 3 & 3 \\ 0 & 1 & 0 \\ 0 & 0 & 1 \end{bmatrix} \begin{bmatrix} 1 & 0 & 0 \\ -1 & 1 & 0 \\ -1 & -1 & 1 \end{bmatrix} \begin{matrix} \\ \\ R1 - 3R3 \end{matrix} \Rightarrow \begin{bmatrix} 1 & 3 & 0 \\ 0 & 1 & 0 \\ 0 & 0 & 1 \end{bmatrix} \begin{bmatrix} 4 & 3 & -3 \\ -1 & 1 & 0 \\ -1 & -1 & 1 \end{bmatrix}$$

Finally we need one more operation to convert the given matrix to a unity matrix, which is shown below.

$$\begin{bmatrix} 1 & 3 & 0 \\ 0 & 1 & 0 \\ 0 & 0 & 1 \end{bmatrix} \begin{bmatrix} 4 & 3 & -3 \\ -1 & 1 & 0 \\ -1 & -1 & 1 \end{bmatrix} R1 - 3R2 \quad \Rightarrow \begin{bmatrix} 1 & 0 & 0 \\ 0 & 1 & 0 \\ 0 & 0 & 1 \end{bmatrix} \begin{bmatrix} 7 & 0 & -3 \\ -1 & 1 & 0 \\ -1 & -1 & 1 \end{bmatrix}$$

Therefore, the inverse of the given matrix $[A] = \begin{bmatrix} 1 & 3 & 3 \\ 1 & 4 & 3 \\ 2 & 7 & 7 \end{bmatrix}$ is $[A]^{-1} = \begin{bmatrix} 7 & 0 & -3 \\ -1 & 1 & 0 \\ -1 & -1 & 1 \end{bmatrix}$.

The Gaussian method outlined here can be extended to any large size matrix and also that it is easily adoptable to write a code for a high speed computer. Since it takes minimum time many computer codes adopt this method for structural analysis.

Besides the Gaussian method, a determinant method can be applied as outlined below.

Determinant Method for Inversion:

For the given non-singular matrix $[A] = \begin{bmatrix} 1 & 3 & 3 \\ 1 & 4 & 3 \\ 2 & 7 & 7 \end{bmatrix}$, the following procedure is adopted.

- replace each element of $[A]$ by its cofactor
- transpose the result to form the adjoint matrix
- then $[A]^{-1} = \frac{1}{|A|} \operatorname{adj}[A]$

First find the determinant of $[A]$

$$|A| = \begin{vmatrix} 1 & 3 & 3 \\ 1 & 4 & 3 \\ 2 & 7 & 7 \end{vmatrix} = 1 \times \begin{vmatrix} 4 & 3 \\ 7 & 7 \end{vmatrix} - 3 \times \begin{vmatrix} 1 & 3 \\ 2 & 7 \end{vmatrix} + 3 \times \begin{vmatrix} 1 & 4 \\ 2 & 7 \end{vmatrix} = 7 - 3 - 3 = 1$$

Replace each element of $[A]$ by its minor

$$[A] = \begin{bmatrix} 1 & 3 & 3 \\ 1 & 4 & 3 \\ 2 & 7 & 7 \end{bmatrix} = \begin{bmatrix} \begin{vmatrix} 4 & 3 \\ 7 & 7 \end{vmatrix} & \begin{vmatrix} 1 & 3 \\ 2 & 7 \end{vmatrix} & \begin{vmatrix} 1 & 4 \\ 2 & 7 \end{vmatrix} \\ \begin{vmatrix} 3 & 3 \\ 7 & 7 \end{vmatrix} & \begin{vmatrix} 1 & 3 \\ 2 & 7 \end{vmatrix} & \begin{vmatrix} 1 & 3 \\ 2 & 7 \end{vmatrix} \\ \begin{vmatrix} 3 & 3 \\ 4 & 3 \end{vmatrix} & \begin{vmatrix} 1 & 3 \\ 1 & 3 \end{vmatrix} & \begin{vmatrix} 1 & 3 \\ 1 & 4 \end{vmatrix} \end{bmatrix} = \begin{bmatrix} 7 & 1 & -1 \\ 0 & 1 & 1 \\ -3 & 0 & 1 \end{bmatrix}$$

Obtain the matrix of cofactors by altering the signs using $\begin{matrix} + & - & + \\ - & + & - \\ + & - & + \end{matrix}$; where a + sign is met no action is taken and where a − sign is met the sign is changed. Cofactor matrix is therefore given by

$$\begin{bmatrix} 7 & -1 & -1 \\ 0 & 1 & -1 \\ -3 & 0 & 1 \end{bmatrix}$$

The adjoint is the transpose of the cofactor matrix.

$$\mathrm{adj}[A] = \begin{bmatrix} 7 & 0 & -3 \\ -1 & 1 & 0 \\ -1 & -1 & 1 \end{bmatrix}$$

The inverse $[A]^{-1} = \frac{1}{|A|}\mathrm{adj}[A]$ is

$$[A]^{-1} = \frac{1}{1} \times \begin{bmatrix} 7 & 0 & -3 \\ -1 & 1 & 0 \\ -1 & -1 & 1 \end{bmatrix} = \begin{bmatrix} 7 & 0 & -3 \\ -1 & 1 & 0 \\ -1 & -1 & 1 \end{bmatrix}$$

Eigen Value Problem:

So far we dealt with linear algebraic equations in matrix form $[A]\{X\} = \{B\}$. We will consider now another class of the linear algebra problem $[A]\{X\} = \lambda\{X\}$, called Eigen Value problem. Augustin-Louis Cauchy used the term *characteristic root* or *eigenvalue* and the equation to determine the Eigen values *characteristic equation*. The steps used earlier to subtract a multiple of one row from another are not valid here because the elimination changes the eigenvalues.

Determinants can be used to give a transition from $[A]\{X\} = \{B\}$ to $[A]\{X\} = \lambda\{X\}$. In these cases the determinant leads to Cramer's rule $\{X\} = [A]^{-1}\{B\}$ and to the characteristic polynomial $|[A] - \lambda[I]|$ whose roots will be Eigen values. Computing λ is more difficult than solving $[A]\{X\} = \{B\}$ particularly when the size is more than 3. Eigenvalues are useful in many Physics phenomena e.g., in Solid Mechanics, the subject matter of this book, Principal stresses and Principal Strains are determined using Eigen Solutions.

Eigenvalue problem $[A]\{X\} = \lambda\{X\}$ involves two unknowns λ and $\{X\}$. The number λ is an *eigenvalue* of the matrix [*A*], and the vector $\{X\}$ is the associated *eigenvector*. As it is $[A]\{X\} = \lambda\{X\}$ is nonlinear since λ multiplies $\{X\}$. If we have λ, then the equation for $\{X\}$ becomes linear. To achieve this, we simply replace $\lambda\{X\}$ on the right-hand side by $\lambda[I]\{X\}$ and rearrange to give

$$[[A] - \lambda[I]]\{X\} = 0$$

The vector $\{X\}$ is said to be in *null space* of $[A] - \lambda[I]$. The number λ is so chosen that $[A] - \lambda[I]$ has a null space. The vector $\{X\} = 0$ always satisfies the Eigenvalue problem above, however we want a *nonzero* eigenvector $\{X\}$. This also implies $[A] - \lambda[I]$ is singular, i.e., the determinant $|[A] - \lambda[I]| = 0$. This is the *characteristic equation.* Expansion of this determinant gives the characteristic polynomial in λ whose roots are the Eigen values. Consider as an example, a 2 × 2 Eigenvalue problem with

$$[A] = \begin{bmatrix} 4 & -5 \\ 2 & -3 \end{bmatrix}$$

The characteristic equation is

$$\begin{vmatrix} 4-\lambda & -5 \\ 2 & -(3+\lambda) \end{vmatrix} = 0$$
$$\Rightarrow -(4-\lambda)(3+\lambda) + 10 = 0$$
$$\Rightarrow \lambda^2 - \lambda - 2 = 0$$

The roots of this polynomial give the Eigenvalues.

$$\lambda_{1,2} = \frac{1 \pm \sqrt{9}}{2} = -1.2$$

A singular matrix must have nonzero vectors [X] in its null space which contains a whole *line* of eigenvectors. For $\lambda_1 = -1$, we have

$$\begin{bmatrix} 4-\lambda & -5 \\ 2 & -(3+\lambda) \end{bmatrix} \begin{Bmatrix} x_1 \\ x_2 \end{Bmatrix} = \begin{vmatrix} 5 & -5 \\ 2 & -2 \end{vmatrix} \begin{Bmatrix} x_1 \\ x_2 \end{Bmatrix} = 0$$

Choosing $x_1 = 1$, we have $2x_2 = 2$. Denoting the Eigen vector for Eigenvalue λ_1 with a subscript 1, we have the first Eigenvector

$$\begin{Bmatrix} x_1 \\ x_2 \end{Bmatrix}_1 = \begin{Bmatrix} 1 \\ 1 \end{Bmatrix}$$

We notice any vector proportional to the above vector also satisfies the Eigenvalue problem, e.g., $\begin{Bmatrix} x_1 \\ x_2 \end{Bmatrix}_1 = \begin{Bmatrix} 1 \\ 1 \end{Bmatrix} = \begin{Bmatrix} 2 \\ 2 \end{Bmatrix} = \begin{Bmatrix} \frac{1}{2} \\ \frac{1}{2} \end{Bmatrix}$... will represent the Eigenvector $\lambda_1 = -1$.

For $\lambda_2 = 2$, we have

$$\begin{bmatrix} 4-\lambda & -5 \\ 2 & -(3+\lambda) \end{bmatrix} \begin{Bmatrix} x_1 \\ x_2 \end{Bmatrix}_2 = \begin{vmatrix} 2 & -5 \\ 2 & -5 \end{vmatrix} \begin{Bmatrix} x_1 \\ x_2 \end{Bmatrix}_2 = 0$$

Choosing once again $x_1 = 1$ (normalizing with respect to x_1), we have $5x_2 = 2$. Denoting the Eigenvector for Eigenvalue λ_2 with a subscript 2, we have the second Eigenvector

$$\begin{Bmatrix} x_1 \\ x_2 \end{Bmatrix}_2 = \begin{Bmatrix} 1 \\ \frac{2}{5} \end{Bmatrix}$$

As above, we can write $\begin{Bmatrix} x_1 \\ x_2 \end{Bmatrix}_2 = \begin{Bmatrix} 1 \\ \frac{2}{5} \end{Bmatrix} = \begin{Bmatrix} 5 \\ 2 \end{Bmatrix}\ldots$ will represent the Eigenvector $\lambda_2 = 2$.

1.2 Vectors and Tensors

Basic sciences are essentially expressing known or observed phenomenon in the form of parameterization to enable a clear understanding of the *state quantities* like displacement velocity, deformation, strain, temperature, pressure, density, current amongst others and their variation in a given medium. All the state quantities are defined at a point in the medium and they can have components in a given direction. The state quantity under question may be isotropic, i.e., a scalar in character or may be the resultant of different components. If we can combine them in accordance to the Parallelogram Law, we have a Vector. It is convenient to express the resultant state vector $\vec{V}$, with three components V_x, V_y, V_z along three orthogonal axes x, y, z. The three axes of such a coordinate system can be righthanded as shown (or can be left handed). Figure 1.9 shows the following vector

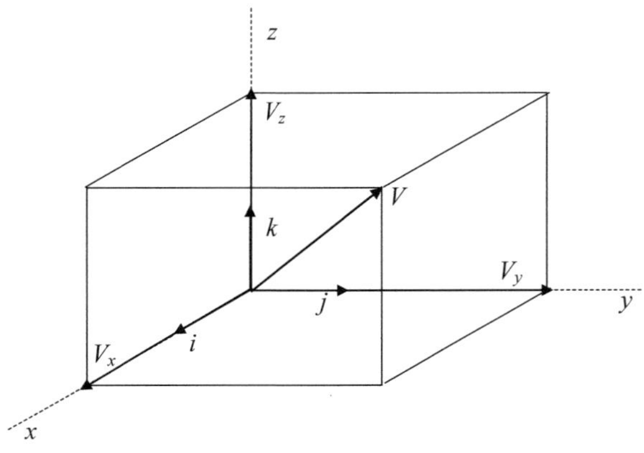

Fig. 1.9 Vector $\vec{V} = V_x\vec{i} + V_y\vec{j} + V_z\vec{k}$

$$\vec{V} = V_x\vec{i} + V_y\vec{j} + V_z\vec{k} \tag{1.1}$$

where $\vec{i}, \vec{j}, \vec{k}$ are unit vectors along x, y and z directions respectively.

The vector in Fig. 1.9 can also be represented in a different frame x', y', z' as in Fig. 1.10. In either of these frames some aspects of the vectors are independent of the reference frame. For example the magnitude of the vector $\vec{V}$ is same. Therefore the sum of the squares of vector components is invariant for a rotation of the reference axes.

$$V_x^2 + V_y^2 + V_z^2 = V_x'^2 + V_y'^2 + V_z'^2 = \vec{V}.\vec{V} \tag{1.2}$$

As we note from Figs. 1.9 and 1.10, the same vector $\vec{V}$ has different components V_x, V_y, V_z in x, y, z reference system and V_x', V_y', V_z' in x', y', z' reference system. As we change the reference system, some of these components may become zero, maximum or minimum etc. which will be of specific interest to a designer. We will see this later, but for the moment we would like to know how the components of a given vector change with transformation of coordinates.

Vector Addition and Subtraction:

Just as a recap, if **A** and **B** (Vectors are also represented by bold letters) are two vectors, their sum or difference is obtained by Parallelogram law. If the sum is a vector **C**, then **C** = **A** + **B** = **B** + **A**. The vector components can also be given by

$$\begin{aligned} A_x + B_x &= C_x \\ A_y + B_y &= C_y \\ A_z + B_z &= C_z \end{aligned} \tag{1.3}$$

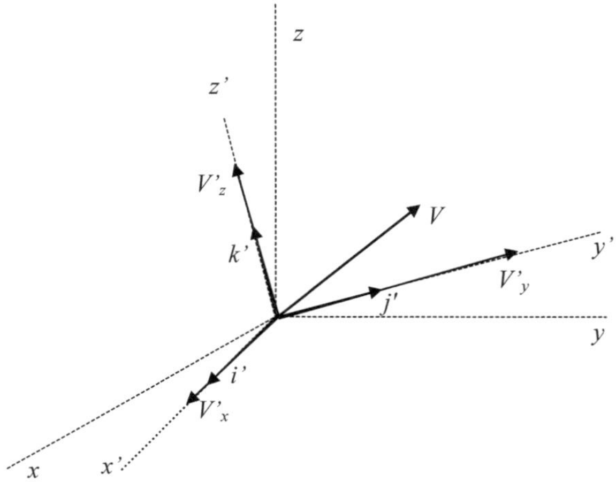

Fig. 1.10 Vector $\vec{V} = V_x'\vec{i}' + V_y'\vec{j}' + V_z'\vec{k}'$

Scalar and Vector Products:

The scalar product of **A** by **B** (or **B** by **A**) is a scalar defined by **A.B** = $AB\cos\theta$ (also called dot product) where θ is the angle between the positive directions of **A** and **B**. The scalar product is thus the product of one vector by the projection of the other vector upon it. Therefore **A.B** = **B.A**. Also

$$\mathbf{i}.\mathbf{i} = \mathbf{j}.\mathbf{j} = \mathbf{k}.\mathbf{k} = 1$$
and
$$\mathbf{i}.\mathbf{j} = \mathbf{j}.\mathbf{k} = \mathbf{k}.\mathbf{i} = 0$$

We can also show that

$$\mathbf{A}.(\mathbf{B} + \mathbf{C}) = \mathbf{A}.\mathbf{B} + \mathbf{A}.\mathbf{C}$$

Similarly, we can show that

$$\mathbf{A}.\mathbf{B} = A_xB_x + A_yB_y + A_zB_z \tag{1.4}$$

The vector product of **A** by **B** (not **B** by **A**) is a vector defined by **A** × **B** = $\boldsymbol{\varepsilon}AB$ sin θ, where $\boldsymbol{\varepsilon}$ is a unit vector perpendicular to the plane of **A** and **B** following righthand rule from the first vector **A** to the second vector **B**. We can express **A** × **B** in terms of its **i**, **j** and **k** components

$$\begin{aligned}\mathbf{A} \times \mathbf{B} &= (\mathbf{i}A_x + \mathbf{j}A_y + \mathbf{k}A_z) \times (\mathbf{i}B_x + \mathbf{j}B_y + \mathbf{k}B_z)\\ &= \mathbf{i} \times \mathbf{i}A_xB_x + \mathbf{i} \times \mathbf{j}A_xB_y + \mathbf{i} \times \mathbf{k}A_xB_z + \mathbf{j} \times \mathbf{i}A_yB_x + \mathbf{j} \times \mathbf{j}A_yB_y + \mathbf{j} \times \mathbf{k}A_yB_z\\ &\quad + \mathbf{k} \times \mathbf{i}A_yB_x + \mathbf{k} \times \mathbf{j}A_yB_y + \mathbf{k} \times \mathbf{k}A_xB_z\\ &= \mathbf{i}(A_yB_z - A_zB_y) + \mathbf{j}(A_zB_x - A_xB_z) + \mathbf{k}(A_xB_y - A_yB_x)\end{aligned}$$

The above can be written as

$$\mathbf{A} \times \mathbf{B} = \begin{vmatrix} i & j & k \\ A_x & A_y & A_z \\ B_x & B_y & B_z \end{vmatrix} \tag{1.5}$$

Coordinate Transformation:

Figure 1.10 shows the vector $\vec{V}$ in both xyz and the primed $x'y'z'$ coordinates. To obtain the primed coordinates in terms of xyz coordinates (transformation of coordinates), we can proceed by projecting the directed line segments x, y, and z, whose vector sum is the directed line segment OV, along the primed coordinate axes. For the x' axis we can get

$$X' = x\cos(x',x) + y\cos(x',y) + z\cos(x',z) \tag{1.6}$$

In the above $\cos(x',x)$ is cosine of the angle between the x' and the x axis etc. We can simplify and use the notation

$$\cos(x',x) = a_{x'x}$$
$$\cos(y',z) = a_{y'z}$$

and rewrite Eq. (1.6) as

$$x' = xa_{x'x} + ya_{x'y} + za_{x'z} \tag{1.6a}$$

a are defined as directional derivatives. Extending (1.6a), we can get the complete coordinate transformation rule as

$$\begin{aligned} x' &= xa_{x'x} + ya_{x'y} + za_{x'z} \\ y' &= xa_{y'x} + ya_{y'y} + za_{y'z} \\ z' &= xa_{z'x} + ya_{z'y} + za_{z'z} \end{aligned} \tag{1.7}$$

The inverse transform can be established by simply changing primed variables to unprimed variables and vice versa. However we maintain the index of a as a primed quantity and maintain the convention as above and interchange the subscripts.

$$\begin{aligned} x &= x'a_{x'x} + y'a_{y'x} + z'a_{z'x} \\ y &= x'a_{x'y} + y'a_{y'y} + z'a_{z'y} \\ z &= x'a_{x'z} + y'a_{y'z} + z'a_{z'z} \end{aligned} \tag{1.8}$$

The notation followed above can be generalized and simplified further. Let us replace x, y, z by x_1, x_2, x_3 and the primed coordinates x', y', z' by x_1', x_2', x_3'. For the subscripts x' and x of the directional derivative a in (1.8) we can use subscript 1; for the subscripts y' and y we can use subscript 2 and similarly for the subscripts z' and z subscript 3. Equations (1.7) and (1.8) in this notation become

$$\begin{aligned} x_1' &= a_{11}x_1 + a_{12}x_2 + a_{13}x_3 \\ x_2' &= a_{21}x_1 + a_{22}x_2 + a_{23}x_3 \\ x_3' &= a_{31}x_1 + a_{32}x_2 + a_{33}x_3 \end{aligned} \tag{1.9a}$$

$$\begin{aligned} x_1 &= a_{11}x_1' + a_{21}x_2' + a_{31}x_3' \\ x_2 &= a_{12}x_1' + a_{22}x_2' + a_{32}x_3' \\ x_3 &= a_{13}x_1' + a_{23}x_2' + a_{33}x_3' \end{aligned} \tag{1.9b}$$

Tensor Notation:

In tensor notation, we usually adopt i, j, k, l or m as subscripts. Whenever these subscripts appear twice as a letter in grouping, we will sum expressions of this grouping with the repeated index taking on values 1, 2, 3 successively. Non-repeated letter or numerical subscripts remain fixed in this process. As an example consider $P_{ij}Q_j$, where j is repeated and i non-repeated. Therefore i remain fixed and j takes repeated values 1, 2 and 3.

$$P_{ij}Q_j = P_{i1}Q_1 + P_{i2}Q_2 + P_{i3}Q_3$$

Under this tensor notation, Eqs. (1.9a) and (1.9b) take a simplified form

$$\begin{aligned} x'_1 &= a_{1j}x_j \\ x'_2 &= a_{2j}x_j \\ x'_3 &= a_{3j}x_j \end{aligned} \tag{1.10a}$$

$$\begin{aligned} x_1 &= a_{j1}x'_j \\ x_2 &= a_{j2}x'_j \\ x_3 &= a_{j3}x'_j \end{aligned} \tag{1.10b}$$

The repeated subscript is also called *Dummy subscript*. The non-repeated index i in $P_{ij}Q_j$, is called *free* or *live* index and such an index represents a set involving the values of 1, 2 and 3. Therefore $P_{ij}Q_j$ represents $\begin{bmatrix} P_{1j}Q_j \\ P_{2j}Q_j \\ P_{3j}Q_j \end{bmatrix}$. Equations (1.10a) and (1.10b) [i.e., (1.9a) and (1.9b)] can be further simplified using tensor notation as follows.

$$\begin{aligned} x'_i &= a_{ij}x_j \\ x_i &= a_{ji}x'_j \end{aligned} \tag{1.11}$$

Free Indices:

Consider A_i; it represents an ordered set of three quantities A_1, A_2 and A_3 and can be represented as a column or a row.

$$A_i = \begin{Bmatrix} A_1 \\ A_2 \\ A_3 \end{Bmatrix}$$

or

$$A_i = \{ A_1 \quad A_2 \quad A_3 \}$$

The vector $\vec{V}$, written alternatively **V** in Fig. 1.10 can be represented by its three scalar components V_1, V_2 and V_3. Therefore $\vec{V}$ or **V** can be written also as V_i.

The number of free indices can be any number in a quantity. In A_{ij} there are free indices, they are a set of nine quantities that includes every permutation of i = 1, 2, 3 and j = 1, 2, 3. We can represent this as an array

$$A_{ij} = \begin{bmatrix} A_{11} & A_{12} & A_{13} \\ A_{21} & A_{22} & A_{23} \\ A_{31} & A_{32} & A_{33} \end{bmatrix}$$

In the above, the first index represents the row and the second index the column in the array. The principal diagonal has repeated indices.

Rotation Matrix:

Now let us consider the rotation of axes; a_{ij} represents a set of nine direction cosines between primed and unprimed axes as in Fig. 1.10. This array is called *rotation matrix*.

$$a_{ij} = \begin{bmatrix} a_{11} & a_{12} & a_{13} \\ a_{21} & a_{22} & a_{23} \\ a_{31} & a_{32} & a_{33} \end{bmatrix}$$

Let us get back to Eqs. (1.10a, 1.10b) dealing with rotation of axes.

$$\begin{aligned} x'_i &= a_{ij}x_j \\ x_i &= a_{ji}x'_j \end{aligned} \tag{1.12a}$$

The second equation in the above can be restated as $x_j = a_{kj}x'_k$ and upon substitution in the first equation, we get

$$x'_i = a_{ij}a_{kj}x'_k \tag{1.13}$$

In the above equation k is a repeated index; when k takes on the numerical value corresponding to the free index i then $a_{ij}a_{kj} = 1$, otherwise when k is different from the free index i $a_{ij}a_{kj} = 0$. Therefore we can state

$$\begin{aligned} a_{ij}a_{kj} &= 0 \quad k \neq i \\ a_{ij}a_{kj} &= 1 \quad k = i \end{aligned} \tag{1.14}$$

Kronecker Delta:

The Kronecker Delta is a symbol defined as

$$\begin{aligned} \delta_{ik} &= 0 \quad k \neq i \\ \delta_{ik} &= 1 \quad k = i \end{aligned}$$

Therefore, from (1.14) we can write

$$a_{ij}a_{kj} = \delta_{ik} \tag{1.15}$$

The Kronecker Delta is widely used in solid or fluid mechanics.

Vector Components:

Let us relook at vector V_i in Fig. 1.10 in unprimed coordinates. Using the Parallelogram law, we can write the components.

$$\begin{aligned} V_x &= V\cos(V, x) \\ V_y &= V\cos(V, y) \\ V_z &= V\cos(V, z) \end{aligned}$$

In primed coordinates, the components can be written down again using the Parallelogram law as

$$\begin{aligned} V'_x &= V_x \cos(x', x) + V_y \cos(x', y) + V_z \cos(x', z) \\ V'_y &= V_x \cos(y', x) + V_y \cos(y', y) + V_z \cos(y', z) \\ V'_z &= V_x \cos(z', x) + V_y \cos(z', y) + V_z \cos(z', z) \end{aligned}$$

Using the double index notation discussed, the vector components in primed coordinates can be written as

$$V'_i = a_{ij}V_j \tag{1.16}$$

Now consider two ordered sets A_i (A_1, A_2, A_3) and B_i (B_1, B_2, B_3) with a reference x_i (x_1, x_2, x_3). In primed axes, these sets are given by

$$\begin{aligned} A'_i &= a_{ij}A_j \\ B'_i &= a_{ij}B_j \end{aligned} \tag{1.16a}$$

Adding the two we have

$$A'_i + B'_i = a_{ij}\left(A_j + B_j\right) \tag{1.17}$$

Tensors, Symmetry and Skew Symmetry:

The transformation equation for a vector A_i is given by Eq. (1.16a)

$$A'_i = a_{ij}A_j$$

The transformation equation above can be generalized; considering a second-order tensor as a set of nine components, therefore with two free subscripts A_{ij}, we can write

$$A'_{ij} = a_{ik}a_{jl}A_{kl} \tag{1.18}$$

In the above a's are familiar direction cosines. The second-order tensor is commonly employed to denote strain and stress at a point in structure and therefore is of significance in Solid Mechanics and other sciences. We can extend these tensors to higher order but will not be dealt with here.

Consider a stress tensor $\boldsymbol{\sigma}$ with direction cosines defined in matrix form $\mathbf{A}$, Eq. (1.18) can be written in the following form to obtain $\boldsymbol{\sigma}'$

$$\boldsymbol{\sigma}' = \mathbf{A}\boldsymbol{\sigma}\mathbf{A}^T$$
$$\begin{bmatrix} \sigma'_{11} & \sigma'_{12} & \sigma'_{13} \\ \sigma'_{21} & \sigma'_{22} & \sigma'_{23} \\ \sigma'_{31} & \sigma'_{32} & \sigma'_{33} \end{bmatrix} = \begin{bmatrix} a_{11} & a_{12} & a_{13} \\ a_{21} & a_{22} & a_{23} \\ a_{31} & a_{32} & a_{33} \end{bmatrix} \begin{bmatrix} \sigma_{11} & \sigma_{12} & \sigma_{13} \\ \sigma_{21} & \sigma_{22} & \sigma_{23} \\ \sigma_{31} & \sigma_{32} & \sigma_{33} \end{bmatrix} \begin{bmatrix} a_{11} & a_{21} & a_{31} \\ a_{12} & a_{22} & a_{32} \\ a_{13} & a_{23} & a_{33} \end{bmatrix} \tag{1.18a}$$

We will use this transformation rule in what follows in the next chapters.

For symmetry of a second-order tensor

$$A_{ij} = A_{ji}$$
$$\Rightarrow$$
$$\begin{bmatrix} A_{11} & A_{12} = A_{21} & A_{13} = A_{31} \\ A_{21} = A_{12} & A_{22} & A_{23} = A_{32} \\ A_{31} = A_{13} & A_{32} = A_{23} & A_{33} \end{bmatrix} \tag{1.19}$$

A skew symmetric tensor of second-order is one where

$$B_{ij} = -B_{ji}$$
$$\Rightarrow$$
$$\begin{bmatrix} 0 & B_{12} & B_{13} \\ -B_{12} & 0 & B_{23} \\ -B_{13} & -B_{23} & 0 \end{bmatrix} \tag{1.20}$$

Note that the diagonal terms in the above are zero. For isotropic materials, the strain and stress tensors that we will deal with are symmetric in general; when subjected to pure shear the tensors are skew-symmetric.

Worked Example 1.1

Show for the three vectors **a**, **b and c** that

$$\mathbf{a}\cdot(\mathbf{b}\times\mathbf{c})=\mathbf{b}\cdot(\mathbf{c}\times\mathbf{a})=\mathbf{c}\cdot(\mathbf{a}\times\mathbf{b})$$

Consider the vector (**a** + **b**) and its cross product with **c**; (**a** + **b**) × **c**. By definition (**a** + **b**) is normal to [(**a** + **b**) × **c**]. Therefore

$$(\mathbf{a}+\mathbf{b})\cdot[(\mathbf{a}+\mathbf{b})\times\mathbf{c}]=0$$

i.e.,

$$\mathbf{a}\cdot(\mathbf{a}\times\mathbf{c})+\mathbf{a}\cdot(\mathbf{b}\times\mathbf{c})+\mathbf{b}\cdot(\mathbf{a}\times\mathbf{c})+\mathbf{b}\cdot(\mathbf{b}\times\mathbf{c})=0$$

The first and last terms in the above are zero, leaving

$$\mathbf{a}\cdot(\mathbf{b}\times\mathbf{c})+\mathbf{b}\cdot(\mathbf{a}\times\mathbf{c})=0,\quad \text{i.e.,}\mathbf{a}\cdot(\mathbf{b}\times\mathbf{c})=\mathbf{b}\cdot(\mathbf{c}\times\mathbf{a})$$

One can show similarly the second part of the problem.

Worked Example 1.2

Three vectors **a**, **b** and **c** form a tetrahedron as shown in Fig. Worked Example 1.2 as shown.

Show that its volume is given by $\frac{1}{6}\left(\vec{a}\times\vec{b}\right).\vec{c}$

$$A_0=\frac{1}{2}\text{base}\times\text{height}=\frac{1}{2}|\vec{a}|\left(\left|\vec{b}\right|\sin\theta\right)=\frac{1}{2}\left|\vec{a}\times\vec{b}\right|$$

We also have the unit vector $\vec{n}=\frac{\vec{a}\times\vec{b}}{\left|\vec{a}\times\vec{b}\right|}$ which is normal to the base, therefore $h_0=\vec{c}.\vec{n}$

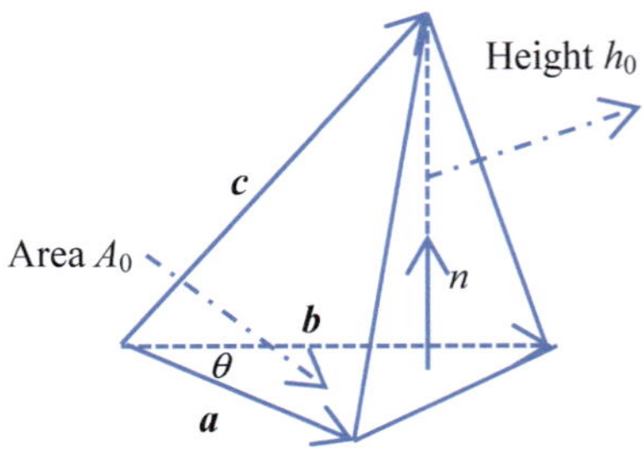

Fig. Worked Example 1.2
Problem Description Figure

$$V_0 = \frac{1}{3} A_0 \times h_0 = \frac{1}{3} \frac{\left|\vec{a} \times \vec{b}\right|}{2} (\vec{c}.\vec{n}) = \frac{1}{6} \left(\vec{a} \times \vec{b}\right) . \vec{c}$$

Worked Example 1.3

$[A]$ and $[B]$ are two 3×3 square matrices, $\{x\}$, $\{y\}$ and $\{z\}$ are 3×1 column matrices in $\{y\} = [A]\{x\} + [B]\{z\}$. Express the matrix equation as a set of scalar equations.

In tensor notation, these equations are

$$y_i = A_{ij} x_j + B_{ij} z_j$$

The free index $i = 1, 2, 3$ and summation over dummy index j is implied.

$$\begin{aligned} y_1 &= A_{11}x_1 + A_{12}x_2 + A_{13}x_3 + B_{11}z_1 + B_{12}z_2 + B_{13}z_3 \\ y_2 &= A_{21}x_1 + A_{22}x_2 + A_{23}x_3 + B_{21}z_1 + B_{22}z_2 + B_{23}z_3 \\ y_3 &= A_{31}x_1 + A_{32}x_2 + A_{33}x_3 + B_{31}z_1 + B_{32}z_2 + B_{33}z_3 \end{aligned}$$

In full matrix form

$$\begin{Bmatrix} y_1 \\ y_2 \\ y_3 \end{Bmatrix} = \begin{bmatrix} A_{11} & A_{12} & A_{13} \\ A_{21} & A_{22} & A_{23} \\ A_{31} & A_{32} & A_{33} \end{bmatrix} \begin{Bmatrix} x_1 \\ x_2 \\ x_3 \end{Bmatrix} + \begin{bmatrix} B_{11} & B_{12} & B_{13} \\ B_{21} & B_{22} & B_{23} \\ B_{31} & B_{32} & B_{33} \end{bmatrix} \begin{Bmatrix} z_1 \\ z_2 \\ z_3 \end{Bmatrix}$$

Worked Example 1.4

[S] is a symmetric matrix and [W] is a skew-symmetric matrix. Determine the scalar value of $S_{ij}W_{ij}$.

Change the dummy index $i \rightarrow p$ and $j \rightarrow q$ and write

$$S_{ij}W_{ij} = S_{pq}W_{pq}$$

Changing the dummy indices from $p \rightarrow j$ and $q \rightarrow i$ we get

$$S_{pq}W_{pq} = S_{ji}W_{ji}$$

Combining, we get

$$S_{ij}W_{ij} = S_{ji}W_{ji}$$

This means effectively both i and j have been changed simultaneously from $i \rightarrow j$ and $j \rightarrow i$. We can also change S_{ij} to S_{ji} since [S] is symmetric and W_{ij} to $-W_{ji}$ since [W] is skew-symmetric.

Therefore

$$S_{ij}W_{ij} = -S_{ij}W_{ij}$$

and it means $S_{ij}W_{ij} = 0$.

Let us consider arbitrarily

$$[\mathrm{S}] = \begin{bmatrix} 2 & 4 & 5 \\ 4 & 3 & 6 \\ 5 & 6 & 1 \end{bmatrix} \quad \text{and} \quad [\mathrm{W}] = \begin{bmatrix} 0 & 2 & 3 \\ -2 & 0 & 4 \\ -3 & -4 & 0 \end{bmatrix}$$

$$\begin{aligned} [\mathrm{S}][\mathrm{W}] &= \begin{bmatrix} 2 & 4 & 5 \\ 4 & 3 & 6 \\ 5 & 6 & 1 \end{bmatrix} \begin{bmatrix} 0 & 2 & 3 \\ -2 & 0 & 4 \\ -3 & -4 & 0 \end{bmatrix} \\ &= \begin{bmatrix} -23 & -16 & 22 \\ -24 & -16 & 24 \\ -15 & 6 & 39 \end{bmatrix} \\ &= -23(-16 \times 39 - 6 \times 24) + 16(-24 \times 39 + 15 \times 24) + 22(-24 \times 6 - 15 \times 16) \\ &= -23(-624 - 144) + 16(-936 + 360) + 22(-144 - 240) \\ &= 23 \times 768 - 16 \times 576 - 22 \times 384 = 17664 - 9216 - 8448 = 0 \end{aligned}$$

Worked Example 1.5

Express any matrix $[A]$ into the sum of a symmetric and a skew-symmetric matrix.

Form a symmetric matrix $[S]$ given by $S_{ij} = \frac{1}{2}\left(A_{ij} + A_{ji}\right)$ and a skew-symmetric matrix $[W]$ given by $W_{ij} = \frac{1}{2}\left(A_{ij} - A_{ji}\right)$.

We can see that $S_{ij} = S_{ji}$, i.e., $[S]$ is symmetric and $W_{ij} = -W_{ji}$, i.e., $[W]$ is skew-symmetric and therefore $S_{ij} + W_{ij} = A_{ij}$.

Worked Example 1.6

The primed axis in Fig. Worked Example 1.6 is obtained by rotating the unprimed axis system through a rotation of angle θ about e_3 axis. Obtain the transformation rule.

$$
\begin{aligned}
e_1' &= e_1 \cos\theta + e_2 \sin\theta + e_3 \times 0 \\
e_2' &= -e_1 \sin\theta + e_2 \cos\theta + e_3 \times 0 \\
e_3' &= e_1 \times 0 + e_2 \times 0 + e_3
\end{aligned}
$$

The transformation matrix is therefore

$$
[Q] = \begin{bmatrix} \cos\theta & \sin\theta & 0 \\ -\sin\theta & \cos\theta & 0 \\ 0 & 0 & 1 \end{bmatrix}
$$

With the help of (1.18), a tensor, $[A]$ expressed in prime coordinates is $[A'] = [Q][A][Q]^T$

$$
\begin{bmatrix} A'_{11} & A'_{12} & A'_{13} \\ A'_{21} & A'_{22} & A'_{23} \\ A'_{31} & A'_{32} & A'_{33} \end{bmatrix} = \begin{bmatrix} \cos\theta & \sin\theta & 0 \\ -\sin\theta & \cos\theta & 0 \\ 0 & 0 & 1 \end{bmatrix} \begin{bmatrix} A_{11} & A_{12} & A_{13} \\ A_{21} & A_{22} & A_{23} \\ A_{31} & A_{32} & A_{33} \end{bmatrix} \begin{bmatrix} \cos\theta & -\sin\theta & 0 \\ \sin\theta & \cos\theta & 0 \\ 0 & 0 & 1 \end{bmatrix}
$$

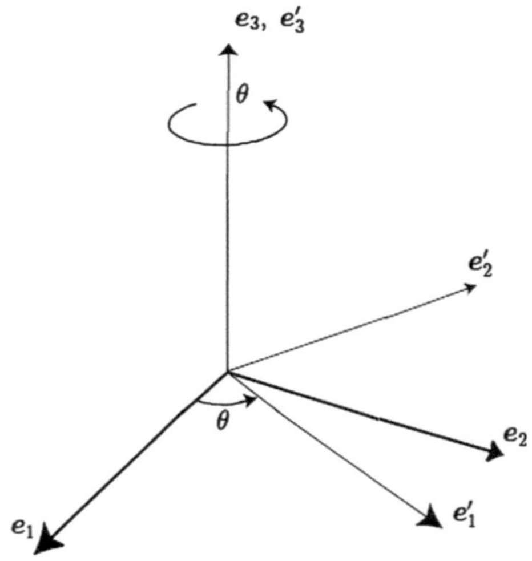

Fig. Worked Example 1.6
Transformation Rule

This gives

$$
\begin{aligned}
A'_{11} &= \frac{1}{2}(A_{11}+A_{22})+\frac{1}{2}(A_{11}-A_{22})\cos 2\theta+\frac{1}{2}(A_{12}+A_{21})\sin 2\theta\\
A'_{12} &= \frac{1}{2}(A_{12}-A_{21})+\frac{1}{2}(A_{12}-A_{21})\cos 2\theta-\frac{1}{2}(A_{11}-A_{22})\sin 2\theta\\
A'_{21} &= -\frac{1}{2}(A_{12}-A_{21})-\frac{1}{2}(A_{12}-A_{21})\cos 2\theta-\frac{1}{2}(A_{11}-A_{22})\sin 2\theta\\
A'_{22} &= \frac{1}{2}(A_{11}+A_{22})-\frac{1}{2}(A_{11}-A_{22})\cos 2\theta-\frac{1}{2}(A_{12}+A_{21})\sin 2\theta\\
A'_{13} &= A_{13}\cos\theta+A_{23}\sin\theta\\
A'_{31} &= A_{31}\cos\theta+A_{32}\sin\theta\\
A'_{23} &= A_{23}\cos\theta-A_{13}\sin\theta\\
A'_{32} &= A_{32}\cos\theta-A_{31}\sin\theta\\
A'_{33} &= A_{33}
\end{aligned}
$$

If $[A]$ is symmetric and $A_{13} = A_{23} = 0$, the above reduce to

$$
\begin{aligned}
A'_{11} &= \frac{1}{2}(A_{11}+A_{22})+\frac{1}{2}(A_{11}-A_{22})\cos 2\theta+A_{12}\sin 2\theta\\
A'_{12} &= -\frac{1}{2}(A_{11}-A_{22})\sin 2\theta\\
A'_{22} &= \frac{1}{2}(A_{11}+A_{22})-\frac{1}{2}(A_{11}-A_{22})\cos 2\theta-A_{12}\sin 2\theta\\
A'_{13} &= 0\\
A'_{23} &= 0\\
A'_{33} &= A_{33}
\end{aligned}
$$

We will learn later that the above are well known Mohr's transformation law in two dimensions.

Derivatives of Vector Quantities:

Let $\mathbf{r} = \boldsymbol{i}x + \boldsymbol{j}y + \boldsymbol{k}z$ represent a vector OA with B nearby on a space curve shown in Fig. 1.11 where $x = x\ (T)$, $y = y\ (T)$, $z = z\ (T)$ and T any real parameter.

$$
\begin{aligned}
\Delta\mathbf{r} &= \overline{AB} = \mathbf{r_1}-\mathbf{r}\\
&= ix_1+jy_1+kz_1-ix-jy-kz\\
&= i(x_1-x)+j(y_1-y)+k(z_1-z)\\
&= i\Delta x+j\Delta y+k\Delta z
\end{aligned}
$$

Then, the derivative of $\mathbf{r}$ with respect to T is a vector that is tangent to the curve at A.

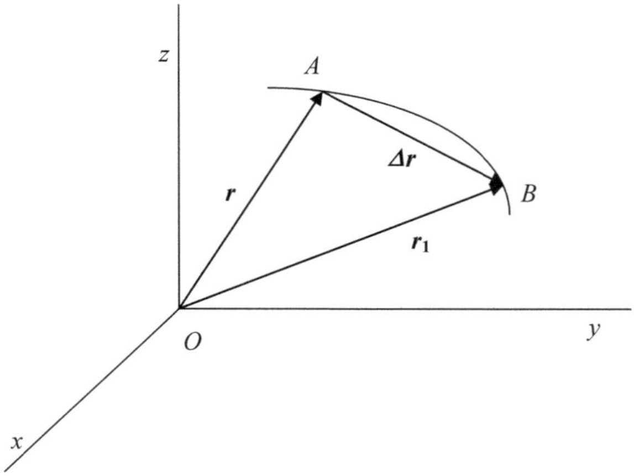

Fig. 1.11 Vector derivative

$$\frac{d\mathbf{r}}{dT} = i\frac{dx}{dT} + j\frac{dy}{dT} + k\frac{dz}{dT} \tag{1.21}$$

Similarly

$$\frac{d^2\mathbf{r}}{dT^2} = i\frac{d^2x}{dT^2} + j\frac{d^2y}{dT^2} + k\frac{d^2z}{dT^2} \tag{1.22}$$

Field Operator, Gradient:

Consider a function ϕ (*x*, *y*, *z*), whose slopes or gradients are given by its partial derivatives with respect to *x*, *y* and *z* respectively in *x*, *y* and *z* directions, $\frac{\partial\phi}{\partial x}, \frac{\partial\phi}{\partial y}, \frac{\partial\phi}{\partial z}$. Using Del operator $\nabla = \frac{\partial}{\partial x}i + \frac{\partial}{\partial y}j + \frac{\partial}{\partial z}k$, the gradient of ϕ is

$$\nabla\phi = \frac{\partial\phi}{\partial x}i + \frac{\partial\phi}{\partial y}j + \frac{\partial\phi}{\partial z}k \tag{1.23}$$

In tensor notation, writing *i*, *j* and *k* unit vectors as i_1, i_2 and i_3 respectively, we have

$$\nabla\phi = \phi_{,j}i_j \tag{1.23a}$$

In the above the subscript comma "," denotes differentiation. The gradient is a vector quantity.

Field Operator Divergence:

The divergence of a vector $\vec{V}$ is a dot product of Del with the vector and denoted by

$$\text{div } \vec{V} = \nabla.\vec{V} = \frac{\partial V_x}{\partial x} + \frac{\partial V_y}{\partial y} + \frac{\partial V_z}{\partial z} \tag{1.24}$$

The divergence of a vector is a scalar as given above. In index notation it is

$$\text{div } \vec{V} = \frac{\partial V_i}{\partial x_i} = V_{i,i} \tag{1.24a}$$

Field Operators Curl:

The next operator is curl, which is a cross product of ∇ with the concerned vector.

$$\text{curl } \vec{V} = \nabla \times \vec{V} = \begin{vmatrix} i & j & k \\ \frac{\partial}{\partial x_1} & \frac{\partial}{\partial x_2} & \frac{\partial}{\partial x_3} \\ V_x & V_y & V_z \end{vmatrix} = \begin{vmatrix} i_1 & i_2 & i_3 \\ \frac{\partial}{\partial x_1} & \frac{\partial}{\partial x_2} & \frac{\partial}{\partial x_3} \\ V_x & V_y & V_z \end{vmatrix} \tag{1.25}$$

The curl is also defined by

$$\nabla \times \vec{V} = i \times \frac{\partial \vec{V}}{\partial x} + j \times \frac{\partial \vec{V}}{\partial y} + k \times \frac{\partial \vec{V}}{\partial z} \tag{1.25a}$$

The field operators, gradient, divergence and curl are all defined with del operator, del of the function, dot product of del with the vector and cross product of del with the vector respectively.

$$\nabla \phi = i\frac{\partial \phi}{\partial x} + j\frac{\partial \phi}{\partial y} + k\frac{\partial \phi}{\partial z} \tag{1.23b}$$

$$\nabla.\phi = \frac{\partial \phi}{\partial x}\frac{\partial \phi}{\partial y}\frac{\partial \phi}{\partial z} \tag{1.24b}$$

$$\nabla \times \phi = \begin{vmatrix} i & j & k \\ \frac{\partial}{\partial x} & \frac{\partial}{\partial y} & \frac{\partial}{\partial z} \\ \phi_x & \phi_y & \phi_z \end{vmatrix} \tag{1.25b}$$

Gauss Theorem:

Gauss theorem enables an integral taken over a volume to be replaced by one taken over the surface bounding that volume, and vice versa. This is a common application while dealing with solids.

Consider a second-order tensor τ_{ij} defined at each point in space in the domain V shown in Fig. 1.12. In the direction x_1 lines parallel to the x_1 axis intersect the boundary only twice. An infintesimal prism of volume δV parallel to x_1 axis with sides dx_2 and dx_3 is shown in the figure. Several such prisms make up the whole volume of the domain V. Over this infinitesimally small prism the integral $\iiint_{\delta V} \frac{\partial \tau_j}{\partial x_1} dx_1 dx_2 dx_3$. Now integrate with respect to x_1 to get

$$\iiint\limits_{\delta V} \frac{\partial \tau_{ij}}{\partial x_1} dx_1 dx_2 dx_3 = \iint \left(\tau_{ij} dx_2 dx_3\right)_R - \iint \left(\tau_{ij} dx_2 dx_3\right)_L \tag{1.26}$$

The first expression on the right-hand side is evaluated at the right end of the prism R, while the second expression is evaluated at the left end of the prism L. Let $\boldsymbol{v}$ represent the unit normal to the boundary of V. v_1, v_2 and v_3 are the direction cosines of $\boldsymbol{v}$. Replace dx_2 and dx_3 at the right end of the prism by $(+v_1 dA)$ and by $(-v_1 dA)$ on the left end of the prism. Then Eq. (1.26) becomes

$$\iiint\limits_{\delta V} \frac{\partial \tau_{ij}}{\partial x_1} dx_1 dx_2 dx_3 = \iint \left(\tau_{ij} v_1 dA\right)_R + \iint \left(\tau_{ij} v_1 dA\right)_L \tag{1.26a}$$

Integrating over all the prisms in V, we have

$$\begin{aligned} \iiint\limits_{V} \frac{\partial \tau_{ij}}{\partial x_1} dV &= \iint\limits_{R} \tau_{ij} v_1 dA + \iint\limits_{L} \tau_{ij} v_1 dA \\ &= \oiint\limits_{S} \tau_{ij} v_1 dA \end{aligned} \tag{1.27}$$

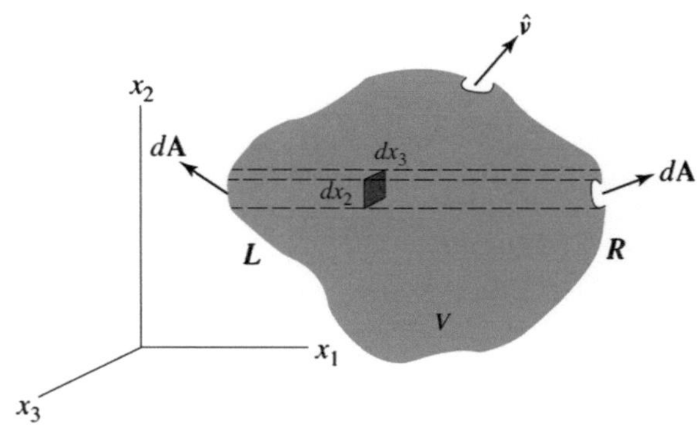

Fig. 1.12 Conversion of volume integral to surface integral

We can generalize the above for any direction x_i and obtain

$$\iiint\limits_V \frac{\partial \tau_{ij}}{\partial x_i} dV = \iiint\limits_V (\tau_{ij})_{,i} dV = \oiint\limits_S \tau_{ij} v_i dA \tag{1.28}$$

where i is a free index.

Divergence Theorem:

If we are dealing with a first order tensor, i.e., a vector V_j Eq. (1.28) becomes

$$\iiint\limits_V V_{j,i} dV = \oiint\limits_S V_j v_i dA \tag{1.29}$$

Performing a contraction operation on the free indices i and j we can write the above

$$\iiint\limits_V V_{j,j} dV = \oiint\limits_S V_j v_j dA$$

or (1.30)

$$\iiint\limits_V \operatorname{div} V dV = \oiint\limits_S V.dA$$

Worked Example 1.7

Determine the gradient of $\phi(\mathbf{x}) = \mathbf{Ax}.\mathbf{x} = A_{ij}x_ix_j$ where $\mathbf{A}$ is a second order tensor.

Calculating the partial derivative of ϕ with respect to x_k

$$\begin{aligned}\phi_{,k} &= A_{ij}(x_ix_j)_{,k} = A_{ij}(x_{i,k}x_j + x_ix_{j,k}) \\ &= A_{ij}(\delta_{ik}x_j + x_i\delta_{jk}) \\ &= A_{kj}x_j + A_{ik}x_i = (A_{kj} + A_{jk})x_j\end{aligned}$$

Or

$$\nabla\phi = (\mathbf{A} + \mathbf{A}^{\mathrm{T}})\mathbf{x}$$

Worked Example 1.8

$\phi(\mathbf{x})$ and $\mathbf{u}(\mathbf{x})$ are a scalar and vector, determine (a) div ($\phi\mathbf{u}$) and (b) grad ($\phi\mathbf{u}$)

$$\begin{aligned}&\text{(a)}\quad \mathrm{div}(\phi\mathbf{u}) = (\phi u_i)_{,i} = u_i\phi_{,i} + \phi u_{i,i}\\&\text{(b)}\quad \mathrm{grad}(\phi\mathbf{u}) = (\phi u_i)_{,j} = u_i\phi_{,j} + \phi u_{i,j}\end{aligned}$$

Worked Example 1.9

$\phi(\mathbf{x})$ and $\mathbf{A}(\mathbf{x})$ are a scalar and second order tensor, determine div ($\phi\mathbf{A}$)

$$\mathrm{div}\,(\phi\mathbf{A}) = \left(\phi A_{ij}\right)_{,j} = A_{ij}\phi_{,j} + \phi A_{ij,j}$$

1.3 Energy Principle

The fundamental principle in Physics is that the energy in the Universe is conserved: it can change in form but cannot be created or destroyed. In elasticity problems a structure deforms utilizing the work done by a load. It deforms in such a way that the deformation (therefore, strained and stressed) of the body absorbs all the work done by external loads through the deformation, not leaving any work (or energy) behind. Since energy cannot be destroyed; the work from external forces or loads is stored in the form of Internal energy.

To determine this condition, we assume a deformation path or pattern in the body, which we can call deformation as a function of the three coordinates in which the body is originally described. If this deformed path or solution satisfies the criteria we stated by bringing a balance between the work input and the stored internal energy, the assumed path is exact. Since we do not know to start with this deflection and its effects such as strain and stress, any assumed path will violate this law. Euler and Lagrange used this principle to determine the condition for exactness by setting up an expression for total energy, in this case coming from straining of the body that is the strain energy stored in the body minus the input work, called total potential energy. This left over total potential energy should be zero, which condition does not yield us a solution. Euler and Lagrange derived one of the most profound laws that this energy can be minimized by varying the approximate solutions that can be assumed. This leads to minimization of the total potential energy.

Here we briefly recall the calculus of a single variable that will set the tone for the Euler-Lagrange equations in Chap. 3 after understanding the deformation, strain and stress in a body on which a load acts. Pierre de Fermat, a 17th century French amateur mathematician who is given credit for early developments that led to infinitesimal calculus, including an original method of finding the greatest and the

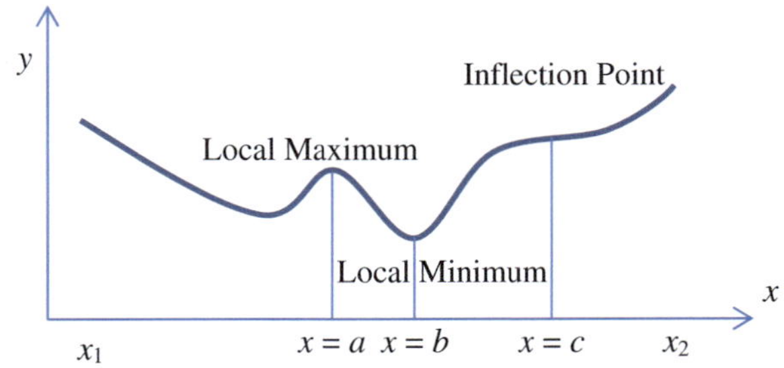

Fig. 1.13 Extremal points of $y = f(x)$

smallest ordinates of curved lines, which is analogous to that of the differential calculus, then unknown.

Consider a function of a single variable, e.g., $y = f(x)$. In this, x is the independent variables and y is a function that depends on the value of x. If in a given range of x, say x_1 to x_2, y is continuous, we know the conditions for this function to have a maximum or a minimum value. For a local maxima, say, at $x = a$, we set the condition that in the neighborhood of $x = a$, y has a value less than $f(a)$. Similarly, for a local minimum at $x = b$, the value of y is more than $f(b)$, in the neighborhood of $x = b$. This is illustrated in Fig. 1.13.

We may call the values of x in the neighborhood of $x = a$ or $x = b$, admissible values for checking the local maxima and minima conditions. To establish the conditions for local maximum or minimum, we expand the function $f(x)$ in Taylor's series about the position $x = a$ or $x = b$. For the region around $x = a$, we get

$$f(x) = f(a) + \left\{\frac{df}{dx}\right\}_{x=a}(x-a) + \frac{1}{2!}\left\{\frac{d^2f}{dx^2}\right\}_{x=a}(x-a)^2 + \frac{1}{3!}\left\{\frac{d^3f}{dx^3}\right\}_{x=a}(x-a)^3 + \cdots$$

This is written using a notation $\frac{df}{dx} = f'(x)$ for the first derivative and $\frac{d^2f}{dx^2} = f''(x)$ for the second derivative etc.

$$f(x) - f(a) + \{f'\}_{x=a}(x-a) + \frac{1}{2!}\{f''\}_{x=a}(x-a)^2 + \cdots \tag{1.31}$$

For $f(a)$ to be a maximum, it is necessary that the above is a negative quantity for all values of x in the neighborhood of a. The quantity $(x - a)$ can be positive or negative, therefore, it is necessary that $f'(a) = 0$.

A similar condition is valid for $f(b)$ to be minimum, i.e., $f'(b) = 0$ from the following.

$$f(x) - f(b) + \{f'(b)\}(x-b) + \frac{1}{2!}\{f''(b)\}(x-b)^2 + \cdots \tag{1.32}$$

Therefore, for $f(a)$ to be a maximum, i.e., $\{f(x) - f(a)\}$ in Eq. (1.31) to be negative, we get the condition that

$$f'(a) = 0 \quad f''(a) > 0 \tag{1.33}$$

Similarly from (1.32), $f(b)$ is minimum if

$$f'(b) = 0 \quad f''(b) > 0 \tag{1.34}$$

We also have a possibility that

$$f'(c) = 0 \quad f''(c) = 0 \tag{1.35}$$

In the above case (1.35) we have neither a maximum nor a minimum for $x = c$, but instead an inflection point as shown in Fig. 1.13. The points satisfying the conditions (1.33)–(1.35) are called extremal points of the function, which can be a maximum, a minimum or an inflection point depending on the above conditions.

The above is a simple case of a function y of a single variable x. The subject becomes interesting when there are more than one variable, for example, the deflection of a plate where there will be two independent variables instead of a beam which we will see to have only one variable as above. In physics, we come across a function of not only the independent variable as above but also the dependent variable and its derivatives such as deformation of a structure. We call such functions as *functionals*; functional is function of a function. When we formulate the problem of a structure we deal with functionals instead of functions alone and it is Euler and Lagrange who made such a formulation to give the most general solution of structures leading to the modern finite element methods of solution with the aid of high performance computing.

1. Exercises

1.1. Determine the inverse of the matrix $[A] = \begin{bmatrix} 0 & 1 & 1 \\ 2 & 3 & -1 \\ -1 & 2 & 1 \end{bmatrix}$ by (a) Gaussian method and (b) determinant method. Also determine its Eigen values and Eigen vectors.

1.2. Show that $\mathbf{A}.(\mathbf{B} \times \mathbf{C}) = \begin{vmatrix} A_x & A_y & A_z \\ B_x & B_y & B_z \\ C_x & C_y & C_z \end{vmatrix}$.

1.3. Show that $\mathbf{A} \times (\mathbf{B} \times \mathbf{C}) = \mathbf{B}(\mathbf{A}.\mathbf{C}) - \mathbf{C}(\mathbf{A}.\mathbf{B})$.

1.4. Derive formulae for $\frac{d}{dT}(\mathbf{P}.\mathbf{Q})$ and $\frac{d}{dT}(\mathbf{P} \times \mathbf{Q})$.

1.5. Obtain the following Operator formulas:

(a) $\nabla.\nabla\phi = \frac{\partial^2\phi}{\partial x^2} + \frac{\partial^2\phi}{\partial y^2} + \frac{\partial^2\phi}{\partial z^2}$

(b) $\nabla \times \nabla\phi = 0$

(c) $\nabla.(\nabla \times \mathbf{F}) = 0$

(d) $\nabla \times (\nabla \times \mathbf{F}) = \nabla(\nabla.\mathbf{F}) - \nabla.\nabla\mathbf{F}$

1.6. [C], [D] and [E] are 3 × 3 square matrices defined in terms of another 3 × 3 square matrices [A] and [B]. Express elements of [C], [D] and [E] matrices in terms of [A] and [B].
1.7. Show that $W_{ij}u_iu_j = 0$ if $[W]$ is skew-symmetric and $\{u\}$ is an arbitrary column matrix.
1.8. Show that the gradient of a constant is zero.
1.9. Determine $\nabla\phi$ if (a) $\phi = xy + xyz$ and (b) $\phi = \log(x + y + z)$ and (c) $\phi = \exp(2x + 2y + 2z)$.
1.10. Show that grad $(\phi\xi)$ = ϕ grad ξ + ξ grad ϕ.
1.11. Determine the directional derivatives of $\phi = x^2yz + 4xz^2$ at the point (1, −2, −1) in the direction of $2\vec{i} - \vec{j} - 2\vec{k}$.
1.12. Determine the direction and magnitude of the maximum of the derivative of $\phi = x^3yz^3$ at the point (2, −1, 1).
1.13. Obtain the directional derivative of $\phi = x^2yz + 4xz^2 + xyz$ at the point (1, 2, 3) in the direction of $2\vec{i} + \vec{j} - \vec{k}$.
1.14. Given $\vec{F} = x^3\vec{i} + y^3\vec{j} + z^3\vec{k}$ determine its divergence and curl.
1.15. Given $\vec{F} = \text{grad}(x^3 + y^3 + z^3 - 3xyz)$ determine its divergence and curl.

Chapter 2
Continuous Solid

We will develop first how a continuous solid will deform under the application of a load. A general theory describing governing equations to understand the solid deformation is *Theory of Elasticity*.

2.1 External and Internal Tractions

The structures discussed above all carry their own weight. Considering a building that we live in as a case the direct loads can be as shown in Fig. 2.1. The direct loads of the roof, floorings, walls etc. are due to gravity here and act vertically down.

There are loads which are imposed on the structure, e.g., the weight of a man as shown in Fig. 2.2 on the structure. This load can be moving also. There can be other imposed loads (or loads acting on the structure), e.g., the wind load as shown in Fig. 2.3. This load is from the pressure of the wind on the roof which therefore is perpendicular to the roof as shown in Fig. 2.3. This is net load shown by integrating the pressures over the surface of the roof.

Another example of imposed loads is a weight being carried on a crane that is being hoisted from one place to another see Fig. 2.4, that we commonly see in a workshop. This load acts on the beam of the crane.

Yet another familiar example is the piston of an internal combustion engine on which we have a pressure of the burnt gases from combustion, see Fig. 2.5. The pressures integrated over the area of the piston give the load or Force as shown.

These forces or loads on a structure cause them to deform and in this process of deformation we determine the effects of deformation (strain and stress) that allow us to design the structure to be safe. The design of the structure to determine the sizes of a chosen material will be the subject matter of ensuing design courses. The design of any structure stationary, rotating or moving in air, water, or in vacuum (space) is dependent on the state of stress, steady or unsteady. We will learn in due

J.S. Rao, *Simulation Based Engineering in Solid Mechanics*,
DOI 10.1007/978-3-319-47614-8_2

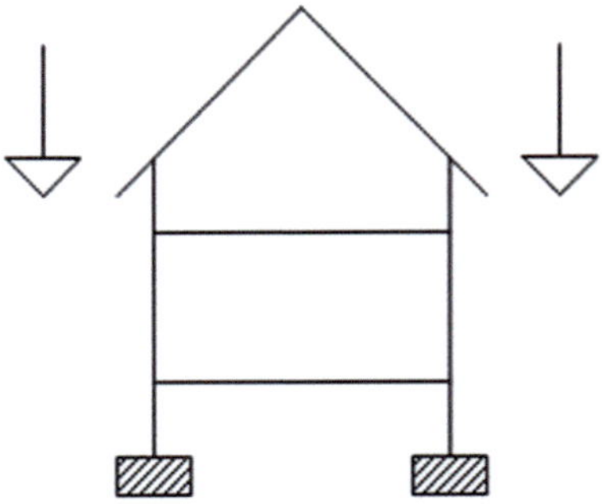

Fig. 2.1 Direct loads

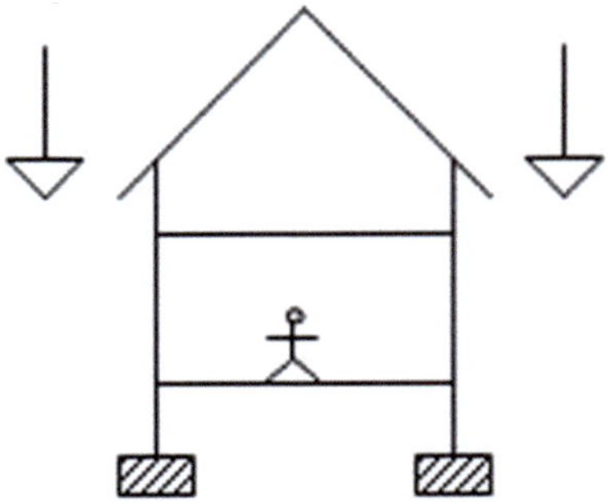

Fig. 2.2 Imposed loads

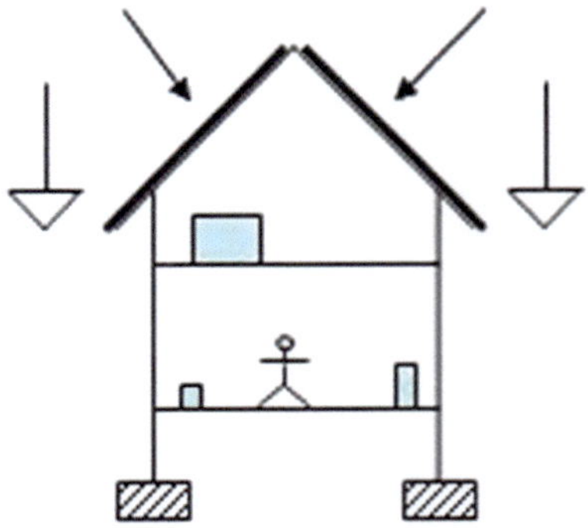

Fig. 2.3 Wind load

course that engineering materials, steel, aluminum, titanium, composites amongst others have all an ultimate stress beyond which they fail or an endurance limit of stress an alternating stress can produce a failure. The stress of course is not same all through a structure, a turbine casing, an alternator rotor, pedestal and foundation, a reciprocating engine casing, an aircraft wing, a railway bridge, a submarine, an off-shore oil rig platform, a space launch vehicle, a naval torpedo etc. amongst others; thus determining stress and where it peaks is the beginning of design of any structure.

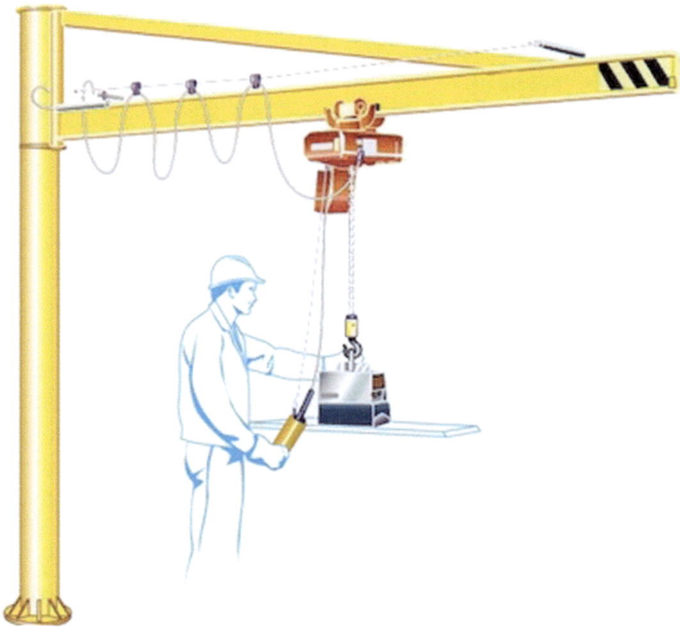

Fig. 2.4 Load imposed on a beam of the crane

Fig. 2.5 Force on the piston

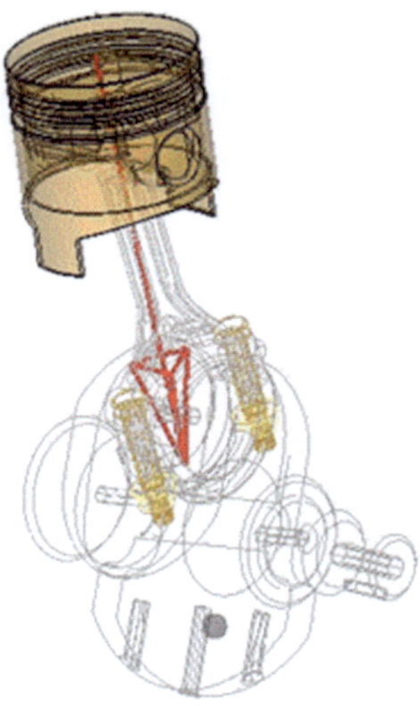

Here we begin with an understanding of a structure, beam, column, plate, shell etc. the forces acting on them and their influence on deformation and internal state of the material of the structure. The deformation, strain and stress vary and therefore we will determine these state quantities at any point in the structure. First we learn how to define in general a force. Euclidean vectors are used to represent physical quantities which have both magnitude and direction, such as force, in contrast to scalar quantities, which have no direction.

The materials mentioned in general are of two types linear or nonlinear; the strain and stress as a result of load can increase linearly with the deformation from the application of loads, then they are subjected linear theory of elasticity. If the behavior of state quantities is not linear, then we have non-linear elasticity. We will confine ourselves with the linear theory of elasticity alone in this book.

Tractions or Loads:

Consider a general solid continuum that would represent a structure as shown in Fig. 2.6 in *XYZ* coordinate system. The *force per unit area* acting at a given location on the body's surface is the *external traction* **T** (bold letters indicate vectors). The elemental surface on the body is denoted by ΔS whose normal is denoted by **ν**. Traction **T** is a *bound vector*, which means **T** cannot slide along its line of action or translate to another location as in a rigid body and keep the same meaning. Notice that *Load* is an engineering definition; actually we should define this as traction because this load cannot be translated. (In engineering we do make exceptions to simplify things). In other words, a traction vector cannot be fully described unless both the force and the surface where the force acts on have been specified. Given both ΔF and ΔS, the traction **T** can be defined as

$$\Delta F = T\Delta S \tag{2.1}$$

It is common to see in everyday life occurrences when structures fail, e.g., buildings collapsing, propeller shafts in a truck failing. The structural components fail because the external loads applied are more than they can carry, generally determined by the state of stress. The material of the structure is said to be failing when internal tractions at some locations are excessive. The internal tractions per unit area inside the structure can exceed a limit (we will define this a little later) at

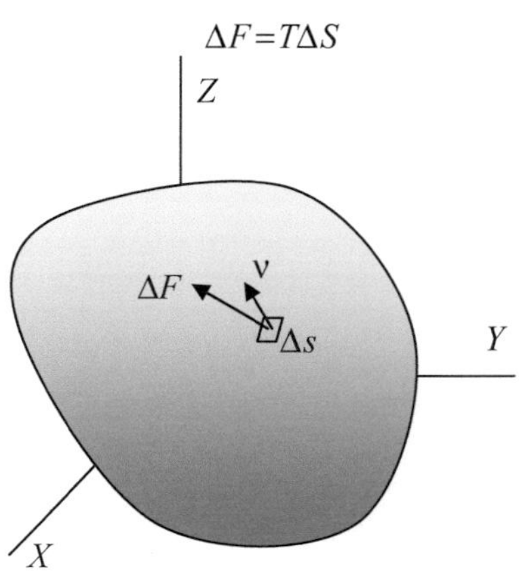

Fig. 2.6 External traction

Fig. 2.7 Internal tractions—stress field

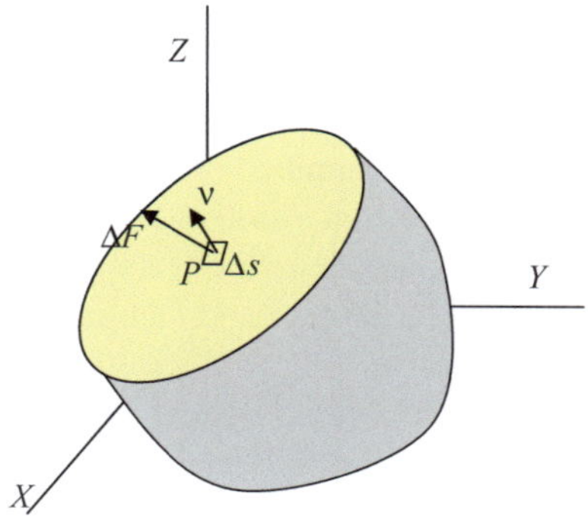

certain locations. We thus define the *stress field* as the distribution of *internal tractions* that balance a given set of *external tractions* and *body forces*. The internal traction within a solid, or *stress*, can be defined in a similar manner.

To define this internal stress let us make an arbitrary slice across the solid as shown in Fig. 2.6, leading to the free body diagram in Fig. 2.7. Surface tractions would appear on the exposed surface, similar in form to the external tractions applied to the body's exterior surface. The stress at point P can be defined using the same equation as was used for **T**.

Stress therefore can be interpreted as internal tractions that act on a defined internal datum plane, in this case the arbitrary slice is made in Fig. 2.7. It should be noted that one cannot measure the stress without first specifying the datum plane.

2.2 Stress Definition

We have seen that stress is a direct consequence of the external traction (or load) that is balanced at any cut cross-section in the body. Let us now define this more specifically.

The word *tensor* was introduced in 1846 by Hamilton. Tensors are geometric objects that describe linear relations between vectors and scalars, and other tensors. Elementary examples include the dot product, the cross product. Vectors and scalars themselves are also tensors. A tensor can be represented as a multidimensional array of numerical values. The order (also degree or rank) of a tensor is the dimensionality of the array needed to represent it, or equivalently, the number of indices needed to label a component of that array. For example, a linear map can be represented by a matrix, a 2-dimensional array, and therefore is a 2nd order tensor. A vector can be represented as a 1-dimensional array and is a 1st order tensor. Scalars are single numbers and are thus Zeroth-order tensors. (Recap yourself with Sect. 1.2 at this time.)

A scalar is described by a single number and a vector with respect to a given basis is described by an array and any tensor with respect to a basis is described by a multidimensional array. The numbers in the array are known as the scalar components of the tensor or simply its components. They are denoted by indices giving their position in the array, in sub and superscript, after the name of the tensor.

The total number of indices required to uniquely specify each component is equal to the dimension of the array, and is called the order or the rank of the tensor. For example, the entries (also called components) of an order 2 tensor **T** would be denoted T_{ij}, where i and j are indices running from 1 to the dimension of the related vector space.

At point P a right-hand coordinate system 123 (xyz) is set as shown in Fig. 2.8a. Now consider a solid element of dimensions $dxdydz$ in the positive directions at this point P. Consider first the datum with a unit normal in direction 1, this face is considered positive if the unit normal is outward, the internal traction on this datum is represented by $\mathbf{T}^{(e1)}$ as shown.

The three components of $\mathbf{T}^{(e1)}$ are denoted by σ_{11}, σ_{21} and σ_{31}. The second subscript denotes the datum surface 1. The first subscript denotes the direction in which the component is pointing and is positive if it is in positive direction on the positive face.

Similarly denote the three components σ_{12}, σ_{22} and σ_{32} on positive surface 2 and σ_{13}, σ_{23} and σ_{33} on positive surface 3 as shown. The negative surfaces on which the outward normals are in the negative directions 1, 2 and 3 can be identified. Positive stresses on these negative surfaces will be directed in the negative directions.

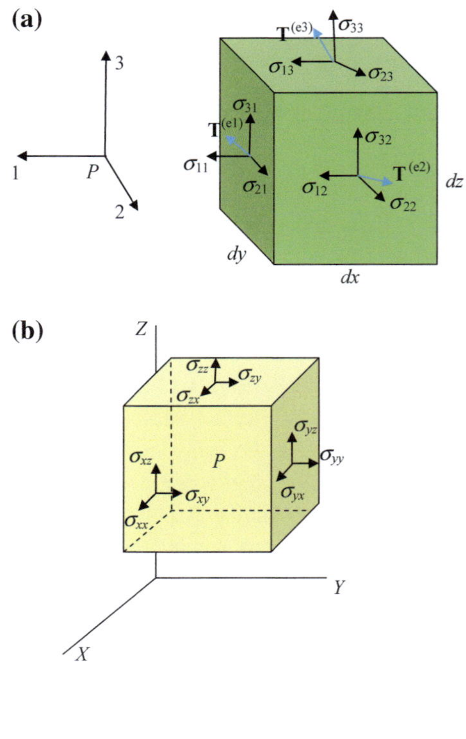

Fig. 2.8 **a** Stress tensor at point P. **b** Stress at point P

Surface tractions, or stresses (usually denoted by σ) acting on an internal datum plane in Fig. 2.7, are taken on an arbitrary slice whose normal is represented by v. Typically we can consider the arbitrary slice taken through the point P to be one of the three planes whose normals are mutually perpendicular and say chosen for convenience along the Cartesian coordinates. We can consider the three mutually perpendicular planes to be forming an infinitely small cube as shown in Fig. 2.8 with the datum planes aligned with Cartesian coordinate system. Remember we are still talking of a location at point P and we use this infinitely small cube to write down the definition of stress at this point. Typically the stress on any plane is decomposed into three mutually orthogonal components given in Fig. 2.8b.

At point P on a chosen datum plane we have three components of the internal traction vector; one component is normal to the surface and represents *direct stress*. The other two components are tangential to the surface and represent *shear stresses*.

We adopt a convenient notation to identify these stress components. First the plane is identified by the normal, e.g., the right-hand side vertical plane has a normal coming out in direction of Y axis, so it is Y surface. If the normal is in the positive direction of Y axis, then it is a positive surface, otherwise the left-hand side vertical face is a negative Y surface since its outward normal is in $-Y$ direction. It also follows that the stress in the negative direction on a negative surface is positive, i.e., tensile.

Every stress component is given two subscripts σ_{ij} the first one i denotes the surface on which it acts. The second subscript j indicates the component direction of the stress; e.g., σ_{yz} indicates the stress component on Y surface acting in Z direction. σ_{xx}, σ_{yy} and σ_{zz} are tensile normal stresses and the rest of the stress components are shear stresses acting in the plane.

What is the distinction between normal and tangential tractions, or equivalently, direct and shear stresses? Direct stresses tend to change the volume of the material (e.g. hydrostatic pressure) and are resisted by the body's bulk modulus (which depends on the Young's modulus and Poisson ratio). Shear stresses tend to deform the material without changing its volume, and are resisted by the body's shear modulus.

The stress state at point P can be represented by an infinitesimal cube with three stress components on each of its six sides (one direct and two shear components). Since each point in the body is under static equilibrium (no net force in the absence of any body forces), only nine stress components from three planes are needed to describe the stress state at a point P. These nine components can be organized into the matrix:

$$\begin{bmatrix} \sigma_{xx} & \sigma_{xy} & \sigma_{xz} \\ \sigma_{yx} & \sigma_{yy} & \sigma_{yz} \\ \sigma_{zx} & \sigma_{zy} & \sigma_{zz} \end{bmatrix} \tag{2.2}$$

where shear stresses across the diagonal are identical ($\sigma_{xy} = \sigma_{yx}$, $\sigma_{yz} = \sigma_{zy}$ and $\sigma_{zx} = \sigma_{xz}$) as a result of static equilibrium (no net moment). This grouping of the nine stress components is known as the *stress tensor* (or stress matrix).

2.3 Equilibrium Relations

We defined stress on an internal datum plane at a point P similar to external traction load applied on the body surface. If we know the external traction i.e., the load applied to the structure, we have to find ways of determining the stress tensor in Eq. (2.2). We have also stated that the body sliced by the internal datum plane is kept in equilibrium by the stress field and the external load. It is expeditious therefore to consider the equilibrium conditions of the infinitesimally small cube in Fig. 2.8.

Now, consider the static equilibrium of the element $dxdydz$ at point P subjected to the body force vector field $\mathbf{b}$. We recall equilibrium conditions, viz.

$$\sum X = 0$$
$$\sum Y = 0$$
$$\sum Z = 0$$

Figure 2.9 shows all the forces in x direction. At point P on the left surface which is negative surface, the stress in the negative direction is positive stress σ_{xx}. This stress varies in all directions, along the x axis it is partial derivative of the stress with respect to x, i.e. $\frac{\partial \sigma_{xx}}{\partial x}$.

Usually, many engineering readers feel, differential equations, particularly partial differential equations is mathematics and of no importance to engineers. It should be noted that differential equations did not arise out of mathematics, but was invented to parameterize Physics and explain the phenomena that occur in Physics. We note here that the physical state quantity, a stress component defined at a point, changes or varies along a given direction. If supposing the stress σ_{xx} under consideration varies only in one direction, say x, we represent this variation of the stress in x direction by $\frac{d\sigma_{xx}}{dx}$, it is just shortening the sentence, the stress σ_{xx} that varies in the direction x, is $\frac{d\sigma_{xx}}{dx}$. It is merely a convenient way of writing this rather stating in a long sentence. When the stress σ_{xx} varies not only in x direction, but also in other

Fig. 2.9 Equilibrium in X direction

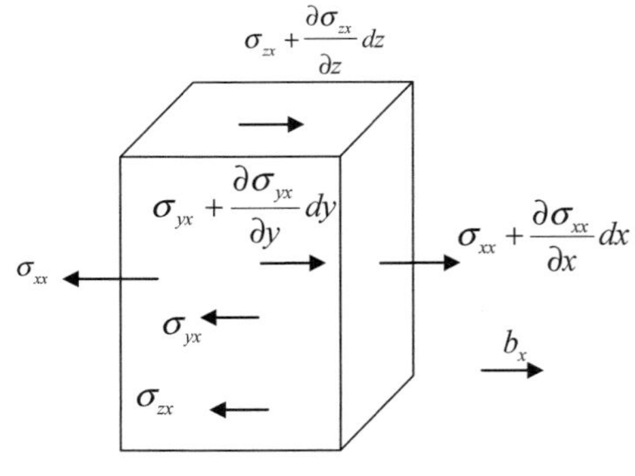

two directions, we denote the derivative as a partial derivative, denoted symbolically $\frac{\partial \sigma_{xx}}{\partial x}$ and read as partial derivative of σ_{xx} with respect to x. Remember this is just a notation at this stage to express our Physics and it is not mathematics so far; for example, do not cancel ∂ in the numerator and denominator, i.e., $\frac{\partial \sigma_{xx}}{\partial x} \neq \frac{\sigma_{xx}}{x}$. The student should familiarize and read in initial stages fully, derivative of the stress in x direction, instead writing mechanically $\frac{d\sigma_{xx}}{dx}$. Now we continue with Physics of the stress and its variation in Fig. 2.9.

Over the distance dx this stress increases by $\frac{\partial \sigma_{xx}}{\partial x} dx$. Therefore the stress on the right-hand surface in positive x direction at distance dx is $\sigma_{xx} + \frac{\partial \sigma_{xx}}{\partial x} dx$ as shown in Fig. 2.9. Likewise the shear stress σ_{zx} on negative Z surface at point P varies along Z axis through a distance dz and σ_{yx}.

Applying equilibrium conditions results in a set of differential equations which govern the stress distribution within the solid.

$$\left(\sigma_{xx} + \frac{\partial \sigma_{xx}}{\partial x} dx\right) dydz + \left(\sigma_{zx} + \frac{\partial \sigma_{zx}}{\partial z} dz\right) dxdy + \left(\sigma_{yx} + \frac{\partial \sigma_{yx}}{\partial y} dy\right) dxdz - \sigma_{xx} dydz - \sigma_{zx} dxdy - \sigma_{yx} dxdz + b_x = 0 \tag{2.3}$$

Writing the equilibrium diagrams in Y and Z directions, forming the equilibrium equations and simplifying we get

$$\begin{aligned} \frac{\partial \sigma_{xx}}{\partial x} + \frac{\partial \sigma_{yx}}{\partial y} + \frac{\partial \sigma_{zx}}{\partial z} + b_x = 0 \\ \frac{\partial \sigma_{xy}}{\partial x} + \frac{\partial \sigma_{yy}}{\partial y} + \frac{\partial \sigma_{zy}}{\partial z} + b_y = 0 \\ \frac{\partial \sigma_{xz}}{\partial x} + \frac{\partial \sigma_{yz}}{\partial y} + \frac{\partial \sigma_{zz}}{\partial z} + b_z = 0 \end{aligned} \tag{2.4}$$

It is advisable to read Eq. (2.4) as partial derivative of stress component σ_{xx} with respect x direction plus partial derivative of stress component σ_{yx} with respect to y direction … rather than noting them as coupled partial differential equations; we are merely expressing the physics in the form of equations. The subject of differential equations grew out of physics; the first partial differential equations were written by Euler while writing the momentum equations of a fluid particle in accordance to Newton's second law.

Now, we have six stress components and only three equations in (2.4); therefore given the body force distribution, the complete stress definition cannot be obtained by solving the above. Obviously we have to bring in additional parameters associated in the deformation of a solid body, one we can quickly think of the deformation of the body itself. We will come to this in the next section. We may note that given the correct stress tensor under a given load application, the stress components will satisfy the above equation.

The concept of equilibrium used above in deriving the general equilibrium equations helps us in determining the stresses, even though we cannot determine the associated quantities like strain and displacements.

Worked Example 2.1

A simple example is the temple columns that have been developed to take the structure weight over a few thousand years ago. Man has learnt that a vertical load can be taken by a tree trunk or stones cut and dressed to do the weight carrying. A simple representation of this structure is shown in Fig. 2.10.

How do we calculate the stress in such a member? Engineering is always about approximation to be able to get the desired answers (or estimates) to allow us proceeding in the design. Several assumptions are made in this process. The six components of stress defined in Eq. (2.2) cannot be obtained from the three coupled partial differential equations in (2.4). Here, we assume that

- The area of cross-section A is very small, i.e., the diameter d is very small compared to the length of the structure in Fig. 2.10.
- By virtue of this we make the assumption that all the stress components except the direct stress $\sigma_{xx} = \sigma$ aligned to the axis of the column are negligible and ignored.

i.e., the simple case of stress is

$$[\sigma] = \begin{bmatrix} \sigma_{xx} & 0 & 0 \\ 0 & 0 & 0 \\ 0 & 0 & 0 \end{bmatrix} \tag{2.5}$$

With this assumption, the net force at any cross-section is σA and should be equal to P. Therefore as an approximation, the longitudinal or axial or direct stress σ is

$$\sigma_{xx} = \sigma = \frac{P}{A} \tag{2.6}$$

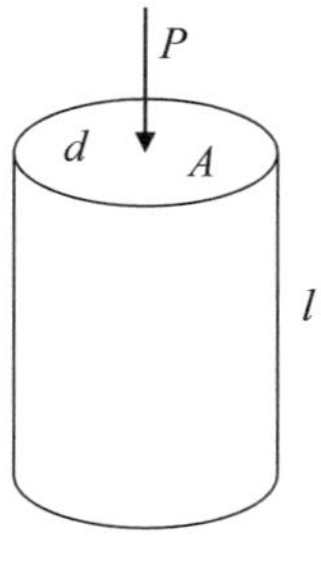

Fig. 2.10 A column taking axial load P

Worked Example 2.2

As another example of an axial load, consider a bolt and nut system, the average stress simplified as one component similar to the above is also given by the same equation in (2.6).

Worked Example 2.3

Consider a structure, the case of a cylindrical vessel of radius r and thickness t subjected to an internal pressure p; e.g., a cooking gas cylinder in our kitchens. The steam generated and stored in a boiler drum in a power plant is also similar to this case. The coordinates used to describe the cylindrical vessel can take advantage of its axial symmetry as given in Fig. 2.11. It is natural to align one coordinate along the axis of the vessel (i.e. in the longitudinal direction). To analyze the stress state in the vessel wall, a second coordinate is then aligned along the circumferential or hoop direction.

A cross-section of the cylindrical vessel is shown in Fig. 2.12.

To keep this section in equilibrium in longitudinal direction with the internal pressure p we need an opposing force that arises from the distribution of load in a direction opposite to the pressure acting on the cylindrical cut surface as shown in the free body diagram Fig. 2.12. This implies that the stress around the wall must have a resultant to balance the internal pressure across the cross-section by a tensile stress σ_l distributed on the cut periphery as shown.

From equilibrium consideration, we can write

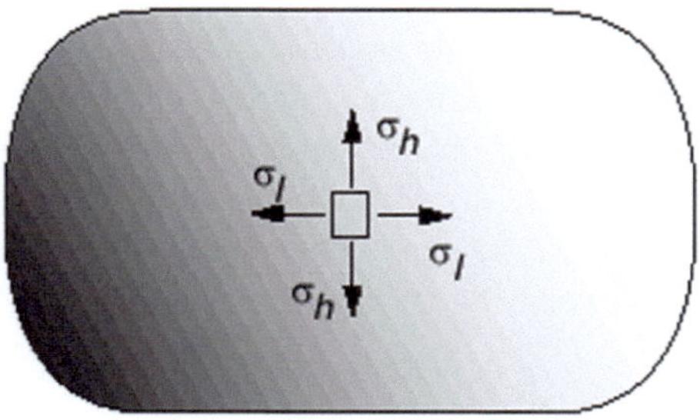

Fig. 2.11 A pressure vessel subjected to internal pressure p

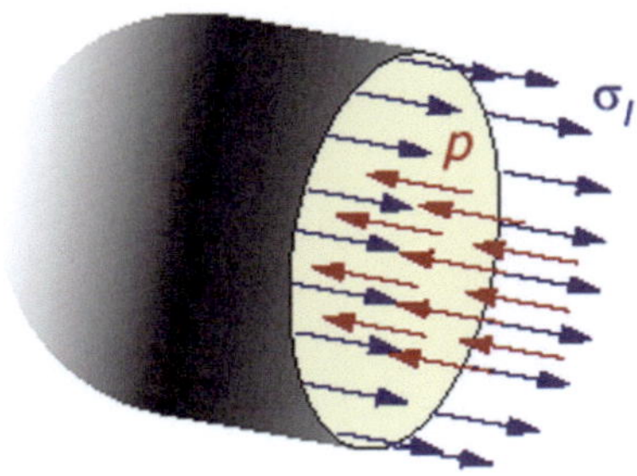

Fig. 2.12 Cross-section of the cylindrical vessel and state of stress

$$\sigma_l t 2\pi r = p\pi r^2 \tag{2.7}$$

i.e.,

$$\sigma_l = \frac{pr}{2t} \tag{2.8}$$

Next we take a longitudinal section of the vessel; assuming the vessel is sufficiently long in this direction, we need to consider only a small section of width dx as shown in Fig. 2.13. The stress on this longitudinal section in the hoop direction is represented by σ_h. Then

$$\begin{aligned} 2\sigma_h t dx &= p2rdx \\ \therefore \sigma_h &= \frac{pr}{t} \end{aligned} \tag{2.9}$$

A solution as above in Examples 2.1–2.3 is an approximate method as followed by our forefathers, e.g., Professor Timoshenko, is called Strength of Materials. This approach depends on the ingenuity of approximation, avoiding Basic Theory of Elasticity that entails detailed stress components determination as propounded in the scientific revolution period. At the dawn of Turbo-machinery from de Laval of Sweden and Charles Parsons in England, the design process became imminent and in absence of computational tools as we have today; the strength of materials provided the means, even if approximate. Because the methods are approximate involving simplified assumptions, the designers over a century ago heavily depended on testing and correction factors. Now with the advent of high speed computers we can directly deal with Theory of Elasticity and obtain highly accurate stress values and improve substantially our design methodologies for high speed and high capacity machines.

The difficulties in the design using the above simplified formulae stems from the fact that the stress at discontinuities, notches … raise considerably from the average values, called stress concentration. Earlier computational methods in the absence of high speed digital computers did not allow us to perform an accurate design but

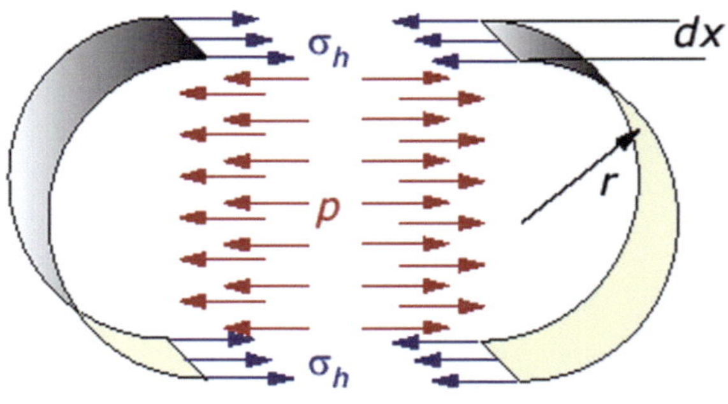

Fig. 2.13 Longitudinal section of the cylindrical vessel and state of stress

Fig. Example 2.4 Pressure Vessel of Fig. 2.11

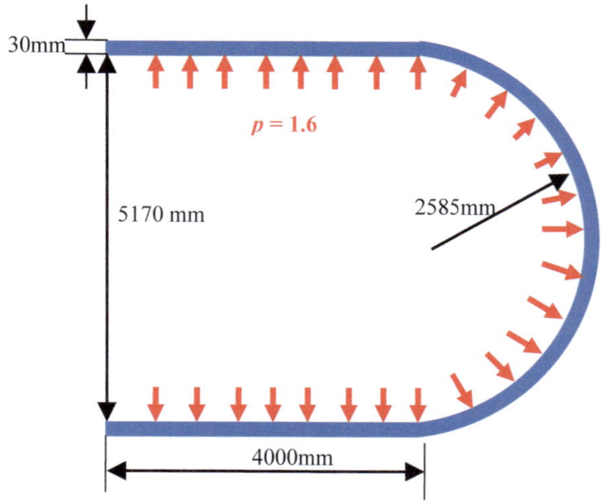

adopt a factor of safety to account for the stress that may increase at these locations. Today's high performance computation changes all that, just the same way the slide rules have been replaced by calculators. We will illustrate this Worked Example 2.3 through modern finite element methods; but first more fundamentals on stress considering two dimensional cases rather than one dimension axial load case.

Worked Example 2.4

Consider the pressure vessel of Fig. 2.11. Let the internal pressure be p = 1.6 MPa, inner diameter d_i = 5170 mm, cylinder length of pressure vessel L = 8000 mm and plate thickness be t = 30 mm (Fig. Example 2.4).

Mean Radius $= \frac{d_i}{2} + \frac{t}{2} = 2.6\,\text{m}$

$$\sigma_l = \frac{pr}{2t} = 69.33\,\text{MPa} \quad \sigma_h = \frac{pr}{t} = 138.667\,\text{MPa}$$

2.3.1 Plane State of Stress

A class of common engineering problems involving stresses in a thin plate or on the free surface of a structural element, such as the surfaces of thin-walled pressure vessels under external or internal pressure, the free surfaces of shafts in torsion and beams under transverse load, have one principal stress (we will define this in the following) that is much smaller than the other two. By assuming that this small principal stress is zero, the three-dimensional stress state can be reduced to two dimensions. Since the remaining two principal stresses lie in a plane, these simplified 2D problems are called plane stress problems, see Fig. 2.14. Note that shear stresses are denoted by τ instead of σ.

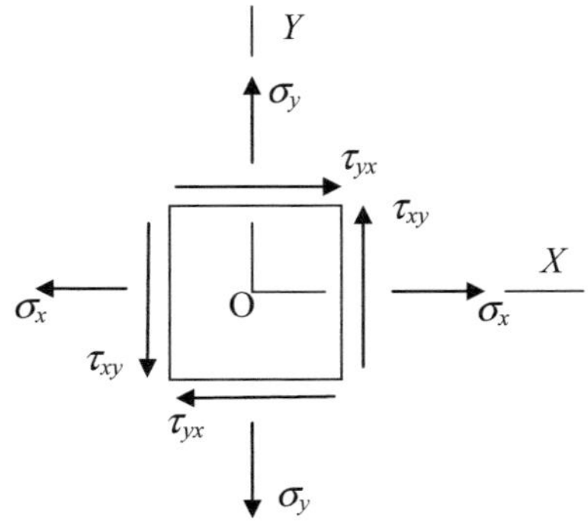

Fig. 2.14 Plane stress

The stress tensor is now 2D and given by

$$\begin{bmatrix} \sigma_x & \tau_{xy} \\ \tau_{yx} & \sigma_y \end{bmatrix} \tag{2.10}$$

Equation (2.4) then reduces to

$$\begin{aligned} \frac{\partial \sigma_{xx}}{\partial x} + \frac{\partial \tau_{yx}}{\partial y} + b_x = 0 \\ \frac{\partial \tau_{xy}}{\partial x} + \frac{\partial \sigma_{yy}}{\partial y} + b_y = 0 \end{aligned} \tag{2.4a}$$

So far we have arbitrarily chosen the datum planes aligned to the Cartesian coordinate system which is also a system in which the solid structure is oriented for analysis. The stress components on these planes tell us their own story; we have no idea where they will be maximum or how they vary if the datum planes are chosen to be different from the Cartesian coordinate system. Stresses in directions that do not line up with the original coordinate set are also important; e.g., the failure plane of a brittle shaft under torsion is often at a 45° angle with respect to the shaft's axis.

2.3.2 *Principal Planes and Principal Stresses*

Consider the coordinate system *XY* in Fig. 2.14 rotated to *X′Y′* through an angle θ with respect to global Cartesian coordinates as in Fig. 2.15. The directional derivatives in Fig. 2.15 are

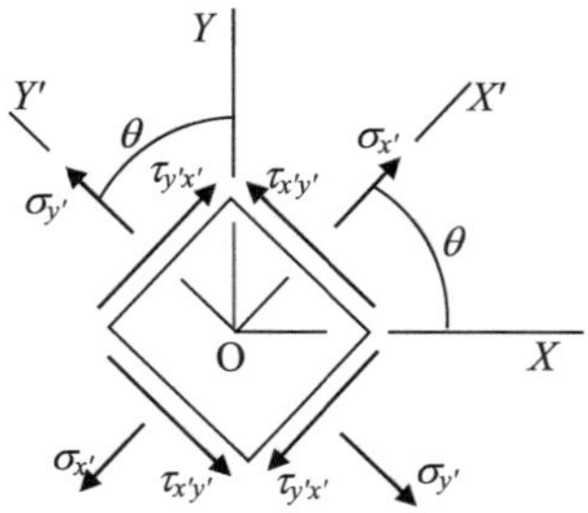

Fig. 2.15 Coordinate transformation

$$a_{11} = \cos X'\widehat{O}X = \cos(-\theta) = \cos\theta$$
$$a_{12} = \cos X'\widehat{O}Y = \cos(90 - \theta) = \sin\theta$$
$$a_{21} = \cos Y'\widehat{O}X = \cos[-(90+\theta)] = \cos[(90+\theta)] = -\sin\theta$$
$$a_{22} = \cos Y'\widehat{O}Y = \cos(-\theta) = \cos\theta$$

$$\mathbf{A} = \begin{bmatrix} a_{11} & a_{12} \\ a_{21} & a_{22} \end{bmatrix} \tag{2.11}$$

The transformation rule for this 2D case from (1.18a) is

$$\begin{gathered} \boldsymbol{\sigma}' = \mathbf{A}\boldsymbol{\sigma}\boldsymbol{\sigma}^{\mathrm{T}} \\ \begin{bmatrix} \sigma'_{11} & \sigma'_{12} \\ \sigma'_{21} & \sigma'_{22} \end{bmatrix} = \begin{bmatrix} a_{11} & a_{12} \\ a_{21} & a_{22} \end{bmatrix} \begin{bmatrix} \sigma_{11} & \sigma_{12} \\ \sigma_{21} & \sigma_{22} \end{bmatrix} \begin{bmatrix} a_{11} & a_{21} \\ a_{12} & a_{22} \end{bmatrix} \end{gathered} \tag{2.12}$$

Substituting, the coordinate transformation above gives the plane stress components in $X'Y'$ coordinates to be

$$\begin{aligned} \sigma_{x'} &= \frac{\sigma_x + \sigma_y}{2} + \frac{\sigma_x - \sigma_y}{2}\cos 2\theta + \tau_{xy}\sin 2\theta \\ \sigma_{y'} &= \frac{\sigma_x + \sigma_y}{2} - \frac{\sigma_x - \sigma_y}{2}\cos 2\theta - \tau_{xy}\sin 2\theta \\ \tau_{x'y'} &= -\frac{\sigma_x - \sigma_y}{2}\sin 2\theta + \tau_{xy}\cos 2\theta \end{aligned} \tag{2.13}$$

Normal stresses $\sigma_{x'}$ and $\sigma_{y'}$ and shear stress $\tau_{x'y'}$ vary smoothly with respect to the rotation angle θ, in accordance with the coordinate transformation Eq. (2.13). We see that there exist a couple of particular angles where the stresses take on special values, Fig. 2.16. First, there exists an angle θ_p where the shear stress $\tau_{x'y'}$ becomes zero. That angle is found by setting $\tau_{x'y'}$ zero in the shear transformation equation of (2.13) and solving for θ (set equal to θ_p). The result is,

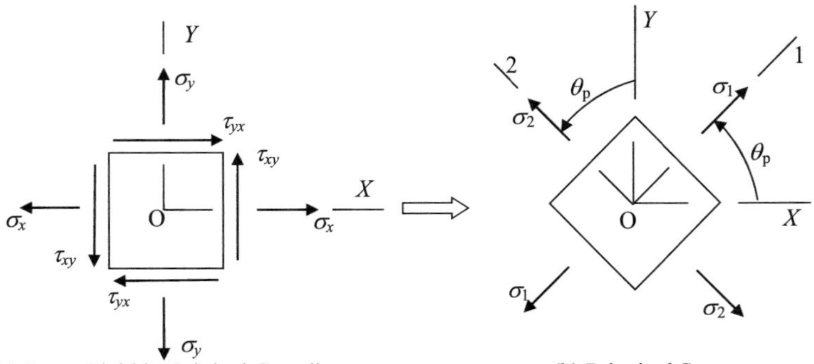

Fig. 2.16 Transformation giving principal directions and principal stresses

$$
\begin{aligned}
-\frac{\sigma_x - \sigma_y}{2} \sin 2\theta + \tau_{xy} \cos 2\theta = 0 \\
\tan 2\theta_p = \frac{2\tau_{xy}}{\sigma_x - \sigma_y}
\end{aligned}
\tag{2.14}
$$

θ_p defines the *principal directions* where the only stresses are normal stresses. These stresses are called *principal stresses* see Fig. 2.16b and are found from the original stresses by substituting $\theta = \theta_p$ from (2.14) in (2.13)

$$
\sigma_{1,2} = \frac{\sigma_x + \sigma_y}{2} \pm \sqrt{\left(\frac{\sigma_x - \sigma_y}{2}\right)^2 + \tau_{xy}^2}
\tag{2.15}
$$

We will learn later the relation between stress and strain determined from *tension tests*. These tests are all uniaxial in nature. The material properties are therefore given by tension tests with tension in one direction; the real structures are however subjected to two dimensional or even three dimensional cases of stresses. If we have only a uniaxial tension test result as a material property, how can we then go forward in designing the structure to be safe? The coordinate transformation comes to our aid in defining the principal stresses as given in (2.15); the designer then ensures the major principal stress not to exceed the ultimate stress the material can withstand. Therefore the importance of determination of principal stresses and strains (similarly maximum shear stress and the plane on which it acts).

We will also learn about controlling stress or equivalent stress derived from distortion energy that will allow us determine the value of stress that can be compared with tension test value. This is commonly known as von Mises stress. This stress is always positive, i.e., tensile. The designs are usually based on this stress criterion.

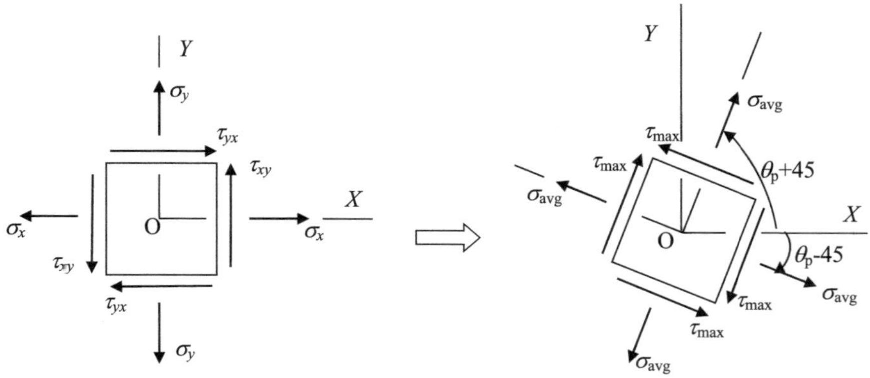

(a) Stress Field in Original Coordinates (b) Maximum Shear Stress

Fig. 2.17 Transformation giving maximum shear stress

2.3.3 *Maximum Shear Stress*

Another important angle is θ_s, where the maximum shear stress occurs. This is determined by finding the maximum of the shear stress transformation equation, and solving for θ. The result is,

$$\begin{aligned} \tan 2\theta_s &= -\frac{\sigma_x - \sigma_y}{2\tau_{xy}} \\ \Rightarrow \theta_s &= \theta_p \pm 45^\circ \end{aligned} \tag{2.16}$$

Substituting for $\theta = \theta_s$ in Eq. (2.13) we get the maximum shear stress to be equal to one-half the difference between the two principal stresses.

$$\begin{aligned} \tau_{max} &= \sqrt{\left(\frac{\sigma_x - \sigma_y}{2}\right)^2 + \tau_{xy}^2} \\ &= \frac{\sigma_1 - \sigma_2}{2} \end{aligned} \tag{2.17}$$

The transformation to the maximum shear stress direction can be illustrated as in Fig. 2.17.

2.3.4 *Mohr's Circle (1867)*

Prior to computer days, most of the designs were limited to simplified 2D cases. Instead of using Eqs. (2.14–2.17) for determining directions of principal planes and principal stresses and maximum shear stresses, engineers were accustomed for simple graphical methods, thus the Mohr's circle approach given here is very popular. The design departments carry out routine activities by an automated approach and engage average technicians to carry out the routine part. Thus fool proof design procedures are standardized and this is where Mohr's graphical approach becomes useful rather than analytical approach. (Even today this is valid, only we use computers as tools rather than skilled engineer scientists; it is the job of engineer-scientist to know what happens inside an analytical or computer program and set up the right process of achieving the results in a mechanical manner.)

We can rewrite the first and third equations in (2.13) as

$$\begin{aligned}\sigma_{x'} - \frac{\sigma_x + \sigma_y}{2} &= \frac{\sigma_x - \sigma_y}{2}\cos 2\theta + \tau_{xy}\sin 2\theta \\ \tau_{x'y'} &= -\frac{\sigma_x - \sigma_y}{2}\sin 2\theta + \tau_{xy}\cos 2\theta\end{aligned} \tag{2.18}$$

Squaring and adding and using $\cos^2 2\theta + \sin^2 2\theta = 1$, we get

$$\left(\sigma_{x'} - \frac{\sigma_x + \sigma_y}{2}\right)^2 + \tau_{x'y'}^2 = \left(\frac{\sigma_x - \sigma_y}{2}\right)^2 + \tau_{xy}^2 \tag{2.19}$$

This is a circle plotted on a graph where the abscissa is the normal stress and the ordinate is the shear stress. This is easier to see if we interpret σ_x and σ_y as being the two principal stresses, and τ_{xy} as being the maximum shear stress. Then we can define the average stress, σ_{avg}, and radius R (which is just equal to the maximum shear stress),

$$\begin{aligned}\sigma_{\text{Avg}} &= \frac{\sigma_x + \sigma_y}{2} \\ R &= \sqrt{\left(\frac{\sigma_x - \sigma_y}{2}\right)^2 + \tau_{xy}^2}\end{aligned} \tag{2.20}$$

The circle is centered at the average stress value, and has a radius R equal to the maximum shear stress, see Fig. 2.18.

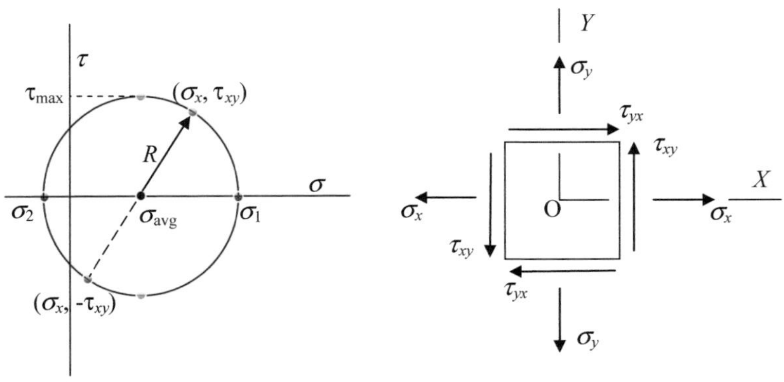

Fig. 2.18 Mohr's circle principal and Max shear stresses

Note that the circle Eq. (2.19) is

$$\left(\sigma_{x'} - \sigma_{\text{Avg}}\right)^2 + \tau_{x'y'}^2 = R^2 \tag{2.19a}$$

The two principal stresses and the maximum shear stress are shown on Mohr's circle. Recall that the normal stresses equal the principal stresses when the stress element is aligned with the principal directions, and the shear stress equals the maximum shear stress when the stress element is rotated 45° away from the principal directions. Also remember from Eq. (2.18), the angle on Mohr's circle is 2θ, see Fig. 2.19. As the stress element is rotated away from the principal (or maximum shear) directions, the normal and shear stress components will always lie on Mohr's Circle. The angle between the current axes (*X* and *Y*) and the principal axes is defined as θ_p, and is equal to one half of the angle between the line *Lxy* and the σ-axis as shown in Fig. 2.20; the procedure to obtain stresses on another plane is illustrated in Fig. 2.20 with the following steps.

1. Locate the two points (σ_x, τ_{xy}) and (σ_y, $-\tau_{xy}$).
2. Draw the line *Lxy* across the circle from (σ_x, τ_{xy}) to (σ_y, $-\tau_{xy}$).
3. Rotate the line *Lxy* by 2θ (twice as much as the angle between *XY* and *X*′*Y*′) and in the *opposite* direction of θ.
4. The stresses in the new coordinates ($\sigma_{x'}\sigma_{y'}$ and $\tau_{x'y'}$) are then read off the circle.

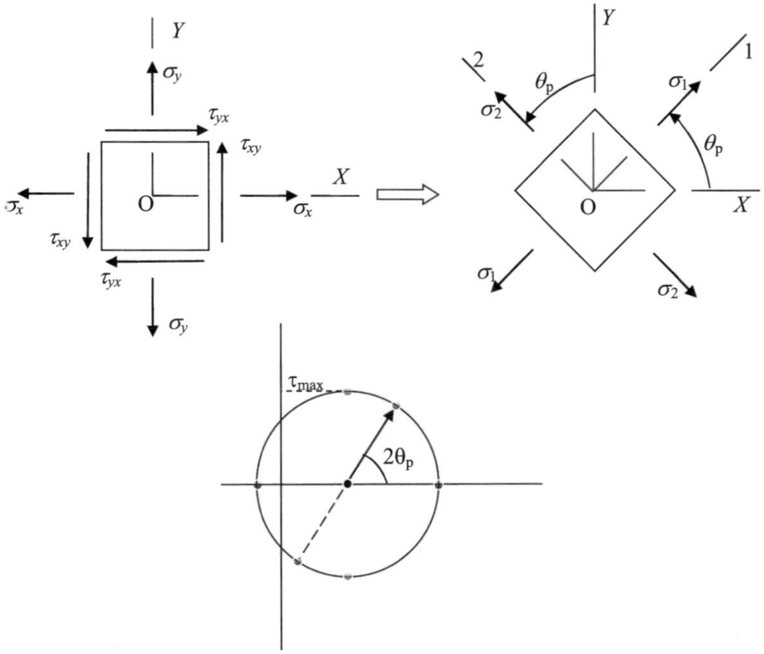

Fig. 2.19 Mohr's circle stress transformation

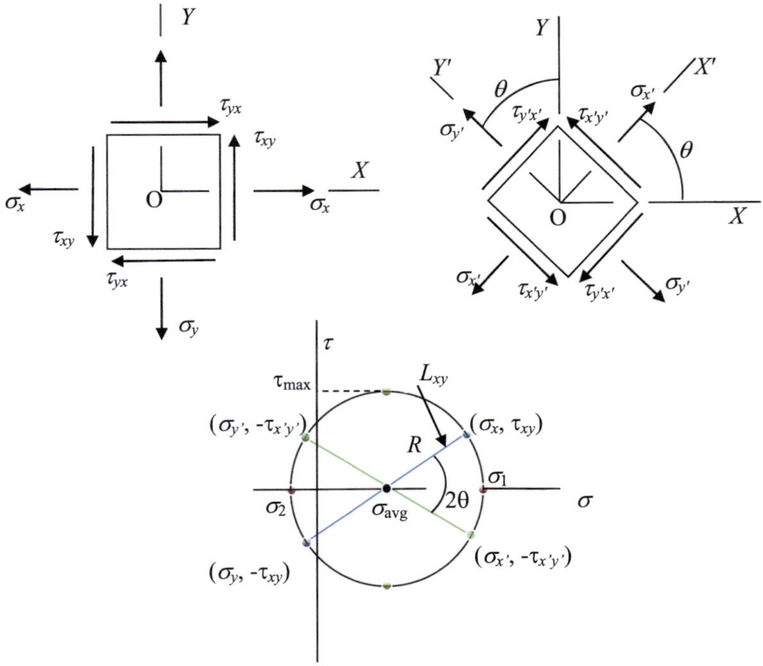

Fig. 2.20 Construction of Mohr's circle for stresses on other planes

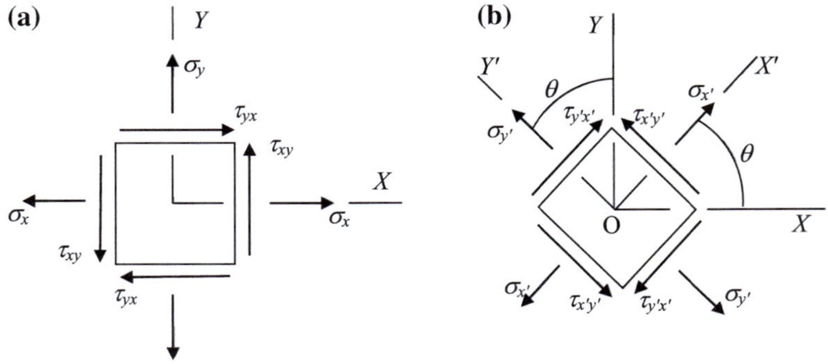

Fig. Worked Example 2.5a, b Plane Stress Problem

Worked Example 2.5

Show that the sum of the normal stresses acting on perpendicular faces for a plane stress element is constant.

Consider the plane stress given in Fig. 2.14 in the unprimed coordinates as reproduced in Fig. Worked Example 2.5a and in the primed coordinates in Fig. Worked Example 2.5b at an angle θ to the unprimed axes.

The stress field in the primed coordinates is obtained from a coordinate transformation as in Eq. (2.13)

$$\sigma_{x'} = \frac{\sigma_x + \sigma_y}{2} + \frac{\sigma_x - \sigma_y}{2}\cos 2\theta + \tau_{xy}\sin 2\theta$$
$$\sigma_{y'} = \frac{\sigma_x + \sigma_y}{2} - \frac{\sigma_x - \sigma_y}{2}\cos 2\theta - \tau_{xy}\sin 2\theta$$
$$\tau_{x'y'} = -\frac{\sigma_x - \sigma_y}{2}\sin 2\theta + \tau_{xy}\cos 2\theta$$

Summing up the first two equations above, we get the desired result.

$$\sigma_{x'} + \sigma_{y'} = \sigma_x + \sigma_y$$

2.3.5 *Triaxial State of Stress—Cauchy Stress Tensor*

The general case of triaxial state of stress is given in Fig. 2.21. The three surfaces with unit normals e_1, e_2 and e_3 and the internal tractions on these three surfaces $\mathbf{T}^{(e1)}$, $\mathbf{T}^{(e2)}$ and $\mathbf{T}^{(e3)}$ respectively are shown in Fig. 2.21. The three components of stresses of these tractions are

$$\boldsymbol{\sigma} = \sigma_{ij} = \begin{bmatrix} \mathbf{T}^{(e1)} \\ \mathbf{T}^{(e2)} \\ \mathbf{T}^{(e3)} \end{bmatrix} = \begin{bmatrix} \sigma_{11} & \sigma_{12} & \sigma_{13z} \\ \sigma_{21} & \sigma_{22} & \sigma_{23} \\ \sigma_{31} & \sigma_{32} & \sigma_{33} \end{bmatrix} \equiv \begin{bmatrix} \sigma_{xx} & \sigma_{xy} & \sigma_{xz} \\ \sigma_{yx} & \sigma_{yy} & \sigma_{yz} \\ \sigma_{zx} & \sigma_{zy} & \sigma_{zz} \end{bmatrix}$$
$$\equiv \begin{bmatrix} \sigma_x & \tau_{xy} & \tau_{xz} \\ \tau_{yx} & \sigma_y & \tau_{yz} \\ \tau_{zx} & \tau_{zy} & \sigma_z \end{bmatrix} \quad (2.2)$$

Using the components of the stress tensor above on surfaces n_i the net internal traction at point P is

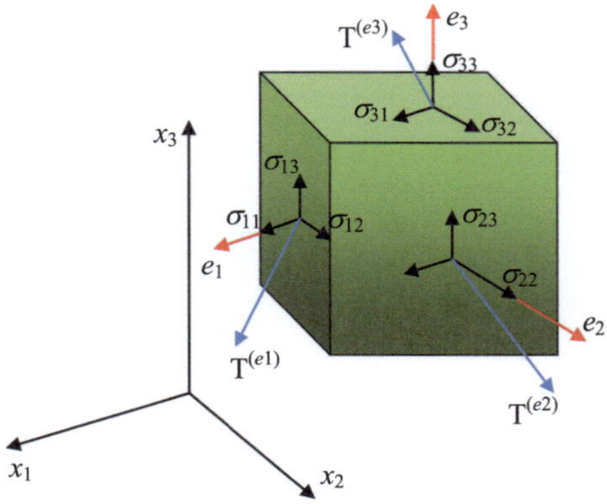

Fig. 2.21 General triaxial stress

$$\begin{aligned}\mathbf{T}^{(\mathbf{n})} &= \mathbf{T}^{(e1)}n_1 + \mathbf{T}^{(e2)}n_2 + \mathbf{T}^{(e3)}n_3 \\ &= \sum_{i=1}^{3}\mathbf{T}^{(ei)}n_i = \left(\sigma_{ij}e_j\right)n_i = \sigma_{ij}n_ie_j\end{aligned} \tag{2.21}$$

Alternatively $T_j^{(n)} = \sigma_{ij}n_i$ gives

$$\begin{bmatrix} T_1^{(n)} & T_2^{(n)} & T3^{(n)} \end{bmatrix} = [n_1 \quad n_2 \quad n_3]\begin{bmatrix} \sigma_{11} & \sigma_{12} & \sigma_{13z} \\ \sigma_{21} & \sigma_{22} & \sigma_{23} \\ \sigma_{31} & \sigma_{32} & \sigma_{33} \end{bmatrix} \tag{2.22}$$

The transformation rule in (1.18a) for three-dimensional case is (Fig. 2.22)

$$\boldsymbol{\sigma}' = \mathbf{A}\boldsymbol{\sigma}\boldsymbol{\sigma}^{\mathrm{T}}$$

$$\begin{bmatrix} \sigma'_{11} & \sigma'_{12} & \sigma'_{13} \\ \sigma'_{21} & \sigma'_{22} & \sigma'_{23} \\ \sigma'_{31} & \sigma'_{32} & \sigma'_{33} \end{bmatrix} = \begin{bmatrix} a_{11} & a_{12} & a_{13} \\ a_{21} & a_{22} & a_{23} \\ a_{31} & a_{32} & a_{33} \end{bmatrix}\begin{bmatrix} \sigma_{11} & \sigma_{12} & \sigma_{13} \\ \sigma_{21} & \sigma_{22} & \sigma_{23} \\ \sigma_{31} & \sigma_{32} & \sigma_{33} \end{bmatrix}\begin{bmatrix} a_{11} & a_{21} & a_{31} \\ a_{12} & a_{22} & a_{32} \\ a_{13} & a_{23} & a_{33} \end{bmatrix} \tag{2.23}$$

Performing the matrix multiplications we can get the individual stress components in the transformed prime coordinate system on the left side in terms of the original Cartesian coordinates where the stress components are given. The direct stresses are

$$\begin{aligned}\sigma'_{11} &= a_{11}^2\sigma_{11} + a_{12}^2\sigma_{22} + a_{13}^2\sigma_{33} + 2a_{11}a_{12}\sigma_{12} + 2a_{11}a_{13}\sigma_{13} + 2a_{12}a_{13}\sigma_{23} \\ \sigma'_{22} &= a_{21}^2\sigma_{11} + a_{22}^2\sigma_{22} + a_{23}^2\sigma_{33} + 2a_{21}a_{22}\sigma_{12} + 2a_{21}a_{23}\sigma_{13} + 2a_{22}a_{23}\sigma_{23} \\ \sigma'_{33} &= a_{31}^2\sigma_{11} + a_{32}^2\sigma_{22} + a_{33}^2\sigma_{33} + 2a_{31}a_{32}\sigma_{12} + 2a_{31}a_{33}\sigma_{13} + 2a_{32}a_{33}\sigma_{23}\end{aligned} \tag{2.24a}$$

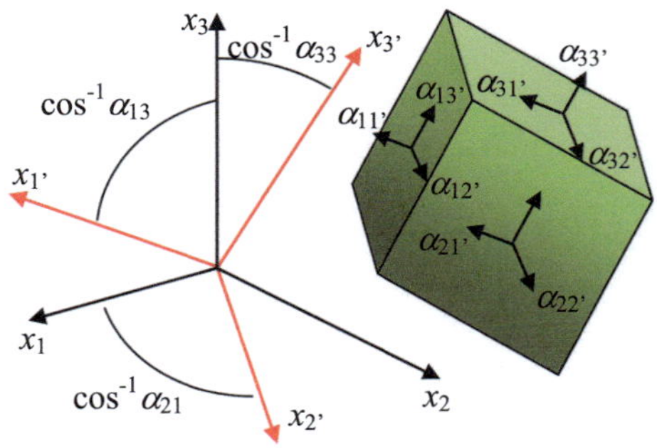

Fig. 2.22 Transformation of stress tensor from *xyz* in Fig. 2.21 to *x′y′z′* axes

The shear stresses in the transformed coordinates are

$$
\begin{aligned}
\sigma'_{12} &= a_{11}a_{21}\sigma_{11} + a_{12}a_{22}\sigma_{22} + a_{13}a_{23}\sigma_{33} + (a_{11}a_{22} + a_{12}a_{21})\sigma_{12} \\
&\quad + (a_{12}a_{23} + a_{13}a_{22})\sigma_{23} + (a_{11}a_{23} + a_{13}a_{21})\sigma_{13} \\
\sigma'_{23} &= a_{21}a_{31}\sigma_{11} + a_{22}a_{32}\sigma_{22} + a_{23}a_{33}\sigma_{33} + (a_{21}a_{32} + a_{22}a_{31})\sigma_{12} \\
&\quad + (a_{22}a_{33} + a_{23}a_{32})\sigma_{23} + (a_{21}a_{33} + a_{23}a_{31})\sigma_{13} \\
\sigma'_{13} &= a_{11}a_{31}\sigma_{11} + a_{12}a_{32}\sigma_{22} + a_{13}a_{33}\sigma_{33} + (a_{11}a_{32} + a_{12}a_{31})\sigma_{12} \\
&\quad + (a_{12}a_{33} + a_{13}a_{32})\sigma_{23} + (a_{11}a_{33} + a_{13}a_{31})\sigma_{13}
\end{aligned}
\tag{2.24b}
$$

As discussed earlier in 2D case, here too at every point in the stressed body there are three principal planes with normal vectors n called principal directions where the corresponding stress vector is perpendicular to the plane, i.e., parallel or in the same direction as the normal vector n, and where there are no shear stresses. The three stresses normal to these principal planes are principal stresses.

The components of the stress tensor depend on the orientation of the coordinate system at the point P under consideration. However, it should be remembered that the stress tensor itself is a physical quantity and it is independent of the coordinate system chosen to represent it. The stress tensor has certain invariants that are also independent of the coordinate system.

2.3.6 Eigen Values and Principal Stresses—General 3D Problem

A stress vector parallel to the normal vector n is given by

$$\mathbf{T}^{(\mathbf{n})} = \lambda\mathbf{n} = \sigma_n\mathbf{n} \tag{2.25}$$

where λ is a constant of proportionality. Here λ corresponds to the magnitudes σ_n of the normal stress vectors which are principal stresses.

Using $T_i^{(n)} = \sigma_{ij}n_j$ with $n_i = \delta_{ij}n_j$, we can write Eq. (2.25) as

$$T_i^{(n)} = \sigma_{ij}n_j = \lambda n_i$$

i.e.,

$$\sigma_{ij}n_j - \lambda n_i = 0 \tag{2.26}$$

or

$$\left(\sigma_{ij} - \lambda\delta_{ij}\right)n_j = 0$$

Equation (2.26) is a homogeneous system, i.e. equal to zero, of three linear equations where n_j are the unknowns. To obtain a nontrivial (i.e., non-zero) solution for n_j, the determinant matrix of the coefficients in (2.26) must be equal to zero, i.e. the system is singular.

$$\left|\sigma_{ij} - \lambda\delta_{ij}\right| = \begin{vmatrix} \sigma_{11} - \lambda & \sigma_{12} & \sigma_{13} \\ \sigma_{21} & \sigma_{22} - \lambda & \sigma_{23} \\ \sigma_{31} & \sigma_{32} & \sigma_{33} - \lambda \end{vmatrix} = 0 \tag{2.27}$$

The above is an Eigenvalue problem and expanding the above gives the characteristic equation. There will be three roots for λ, they are the Eigenvalues also called Characteristic values, λ_1, λ_2 and λ_3. Associated with each of these Eigen values, we have the vectors n_j; they are called Eigenvectors or characteristic vectors.

$$\left|\sigma_{ij} - \lambda\delta_{ij}\right| = \begin{vmatrix} \sigma_{11} - \lambda & \sigma_{12} & \sigma_{13} \\ \sigma_{21} & \sigma_{22} - \lambda & \sigma_{23} \\ \sigma_{31} & \sigma_{32} & \sigma_{33} - \lambda \end{vmatrix} = -\lambda^3 + I_1\lambda^2 - I_2\lambda + I_3 = 0 \tag{2.28}$$

where

$$\begin{aligned} I_1 &= \sigma_{11} + \sigma_{22} + \sigma_{33} = \sigma_{kk} \\ I_2 &= \begin{vmatrix} \sigma_{22} & \sigma_{23} \\ \sigma_{32} & \sigma_{33} \end{vmatrix} + \begin{vmatrix} \sigma_{11} & \sigma_{13} \\ \sigma_{31} & \sigma_{33} \end{vmatrix} + \begin{vmatrix} \sigma_{11} & \sigma_{12} \\ \sigma_{21} & \sigma_{22} \end{vmatrix} \\ &= \sigma_{11}\sigma_{22} + \sigma_{22}\sigma_{33} + \sigma_{11}\sigma_{33} - \sigma_{12}^2 - \sigma_{23}^2 - \sigma_{13}^2 = \frac{1}{2}\left(\sigma_{ii}\sigma_{jj} - \sigma_{ij}\sigma_{ji}\right) \\ I_3 &= \det\left(\sigma_{ij}\right) = \sigma_{11}\sigma_{22}\sigma_{33} + 2\sigma_{12}\sigma_{23}\sigma_{31} - \sigma_{12}^2\sigma_{33} - \sigma_{23}^2\sigma_{11} - \sigma_{13}^2\sigma_{22} \end{aligned} \tag{2.29}$$

The characteristic Eq. (2.28) has three real roots λ_i, i.e. not imaginary due to the symmetry of the stress tensor. The σ_1 = max ($\lambda_1, \lambda_2, \lambda_3$), σ_3 = min ($\lambda_1, \lambda_2, \lambda_3$) and $\sigma_2 = I_1 - \sigma_1 - \sigma_3$ are the principal stresses, functions of the eigenvalues λ_i. The principal stresses are unique for a given stress tensor. Therefore, from the characteristic equation, the coefficients I_1, I_2 and I_3, called the first, second, and third stress invariants, respectively, have always the same value regardless of the coordinate system's orientation.

For each eigenvalue, there is a non-trivial solution for n_j in the equation $\left(\sigma_{ij} - \lambda\delta_{ij}\right)n_j = 0$. These solutions are the principal directions or eigenvectors defining the plane where the principal stresses act. The principal stresses and principal directions characterize the stress at a point and are independent of the orientation.

The stress invariants in terms of principal stresses reduce to

$$\begin{aligned} I_1 &= \sigma_1 + \sigma_2 + \sigma_3 \\ I_2 &= \sigma_1\sigma_2 + \sigma_1\sigma_3 + \sigma_2\sigma_3 \\ I_3 &= \sigma_1\sigma_2\sigma_3 \end{aligned} \tag{2.30}$$

Note there is no Mohr's circle solution for the 3D case; we have to follow the above procedure. The maximum shear stress is obtained by considering the maximum of the three maximum shear stresses of the three principal planes

$$\begin{aligned} \tau_{\max,3} &= \pm\frac{1}{2}(\sigma_1 - \sigma_2) \\ \tau_{\max,2} &= \pm\frac{1}{2}(\sigma_1 - \sigma_3) \\ \tau_{\max,1} &= \pm\frac{1}{2}(\sigma_2 - \sigma_3) \end{aligned} \tag{2.31}$$

This is where modern computers help us in finding the Eigenvalues, i.e., Principal stresses of a complex three-dimensional problem.

Worked Example 2.6

Let the given stress tensor be $\begin{bmatrix} -10 & 10 & 0 \\ 10 & 10 & 0 \\ 0 & 0 & 0 \end{bmatrix}$ MPa for which we determine the principal stresses and directions. First the stress invariants from (2.29) are

$$I_1 = \sigma_{11} + \sigma_{22} + \sigma_{33} = \sigma_{kk} = 0$$

$$I_2 = \begin{vmatrix} \sigma_{22} & \sigma_{23} \\ \sigma_{32} & \sigma_{33} \end{vmatrix} + \begin{vmatrix} \sigma_{11} & \sigma_{13} \\ \sigma_{31} & \sigma_{33} \end{vmatrix} + \begin{vmatrix} \sigma_{11} & \sigma_{12} \\ \sigma_{21} & \sigma_{22} \end{vmatrix} = 0 + 0 + \begin{vmatrix} -10 & 10 \\ 10 & 10 \end{vmatrix} = -200\ (\text{MPa})^2$$

$$I_3 = \det(\sigma_{ij}) = \sigma_{11}\sigma_{22}\sigma_{33} + 2\sigma_{12}\sigma_{23}\sigma_{31} - \sigma_{12}^2\sigma_{33} - \sigma_{23}^2\sigma_{11} - \sigma_{13}^2\sigma_{22} = 0$$

From (2.30)

$$\begin{aligned} &\sigma_1 + \sigma_2 + \sigma_3 = 0 \\ &\sigma_1\sigma_2 + \sigma_1\sigma_3 + \sigma_2\sigma_3 = -200 \\ &\sigma_1\sigma_2\sigma_3 = 0 \end{aligned}$$

It is clear from the third equation above; one of the principal stresses is zero. Since the given stress is two-dimensional in xy plane, $\sigma_3 = 0$.

$$\therefore \sigma_1 + \sigma_2 = 0 \quad \text{and} \quad \sigma_1\sigma_2 = -200$$

$$\sigma_1 + \frac{-200}{\sigma_1} = 0$$

$$\sigma_1 = 14.1421 \text{ MPa} \quad \text{and} \quad \sigma_2 = -14.1421 \text{ MPa}$$

From (2.15) treating the problem as two-dimensional, we have

$$\begin{aligned}
\sigma_{1,2} &= \frac{\sigma_x + \sigma_y}{2} \pm \sqrt{\left(\frac{\sigma_x - \sigma_y}{2}\right)^2 + \tau_{xy}^2} \\
&= \frac{-10 + 10}{2} \pm \sqrt{\left(\frac{-10 - 10}{2}\right)^2 + 10^2} \\
&= \pm\sqrt{100 + 100} = \pm\sqrt{200} = \pm 14.1421 \text{ MPa}
\end{aligned}$$

From (2.14) we have

$$\tan 2\theta_p = \frac{2\tau_{xy}}{\sigma_x - \sigma_y} = \frac{20}{-20} \quad \therefore \theta_p = -22.5^\circ$$

Now we can also verify Eq. (2.20) for the Mohr's circle center and radius and the locations of principal planes as given above.

Equations (2.16) and (2.17) give the maximum shear stress and the planes on which it acts.

$$\tau_{\max} = \frac{\sigma_1 - \sigma_2}{2} = 14.1421 \text{ MPa}$$

$$\theta_s = 22.5^\circ \text{ and } -67.5^\circ$$

2.4 Strain

So far we discussed equilibrium of the body under the applied traction loads and defined the state of stress at a given point. Considering the equilibrium alone led to three coupled partial differential equations with six unknowns. Since the body deforms under the applied traction loads the stress field defined should be connected to the deformations. A given body cannot deform independently to any deformed position under a given traction load; the deformation has to be unique.

While considering the equilibrium state of a solid under external traction we have totally ignored the movement of different points in the structure, e.g., the location where the load is applied itself moves. But these displacements from the unloaded to loaded conditions are so small; they have no influence while writing the equilibrium diagrams.

But the displacements themselves are significant in producing overall deformations in the body. For example consider a rod with initial length L which is

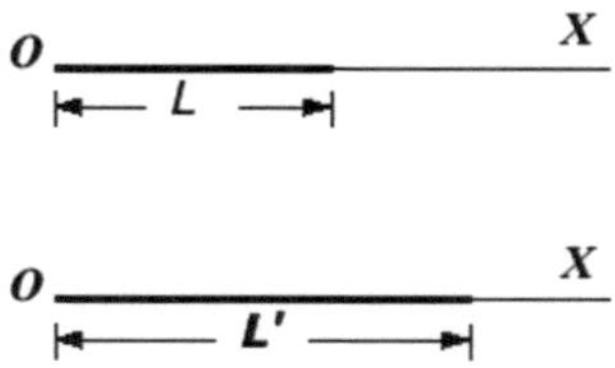

Fig. 2.23 Global strain in a rod

stretched to a length L' as shown in Fig. 2.23. The strain measure e, a dimensionless ratio, is defined as the ratio of elongation with respect to the original length, this is global strain.

$$e = \frac{L' - L}{L} \tag{2.32}$$

2.4.1 Strain at a Point

The strain at each point may vary dramatically if the bar's elastic modulus or cross-sectional area changes. To track down the strain at each point, consider an arbitrary point in the bar P, which has a position vector x, and its infinitesimal neighbor dx. Point P shifts to $\boldsymbol{P}'$, which has a position vector $\boldsymbol{x}'$, after the stretch, see Fig. 2.24. In the meantime, the small "step" dx is stretched to $\boldsymbol{dx}'$. The strain at point P can be defined the same way as in the global strain measure, and obtain the expression for the strain in differential form.

$$\varepsilon = \frac{dx' - dx}{dx} = \frac{du}{dx} \tag{2.33}$$

Now let us consider a body located in Cartesian coordinates XYZ as shown in Fig. 2.25. The unreformed and deformed shapes are shown. During deformation, let point P at position vector R move to P' with the position vector R'. Then the deformation at P is

$$u = R' - R \tag{2.34}$$

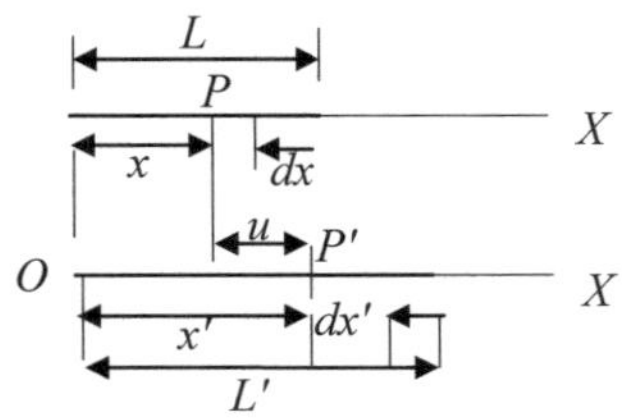

Fig. 2.24 Strain at a point

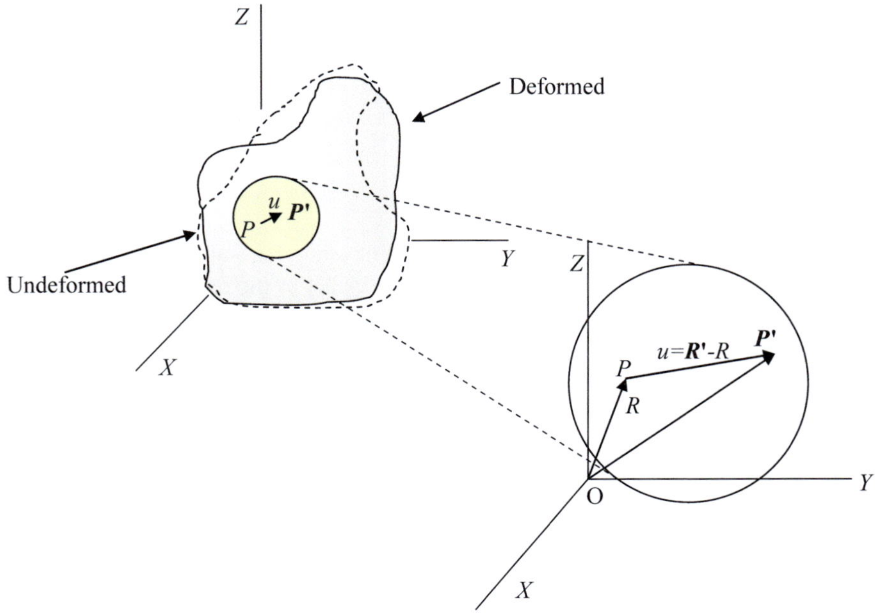

Fig. 2.25 Representation of general 3D strain

The deformed vector $\boldsymbol{u}$ has three components u, v, w and the rate of change of these displacements with x, y and z give us the strains represented by the following strain tensor.

$$\begin{bmatrix} \varepsilon_{xx} & \varepsilon_{xy} & \varepsilon_{xz} \\ \varepsilon_{yx} & \varepsilon_{yy} & \varepsilon_{yz} \\ \varepsilon_{zx} & \varepsilon_{zy} & \varepsilon_{zz} \end{bmatrix} \tag{2.35}$$

These strains in tensor form are given by

$$\varepsilon_{ij} = \frac{1}{2}\left(\frac{\partial u_i}{\partial x_j} + \frac{\partial u_j}{\partial x_i}\right) \tag{2.36}$$

which in expanded form give the normal and shear strains.

$$\begin{aligned} \varepsilon_{xx} &= \frac{\partial u}{\partial x} \quad \varepsilon_{yz} = \frac{1}{2}\left(\frac{\partial v}{\partial z} + \frac{\partial w}{\partial y}\right) = \varepsilon_{zy} \\ \varepsilon_{yy} &= \frac{\partial v}{\partial y} \quad \varepsilon_{zx} = \frac{1}{2}\left(\frac{\partial w}{\partial x} + \frac{\partial u}{\partial z}\right) = \varepsilon_{xz} \\ \varepsilon_{zz} &= \frac{\partial w}{\partial z} \quad \varepsilon_{xy} = \frac{1}{2}\left(\frac{\partial u}{\partial y} + \frac{\partial v}{\partial x}\right) = \varepsilon_{yx} \end{aligned} \tag{2.37}$$

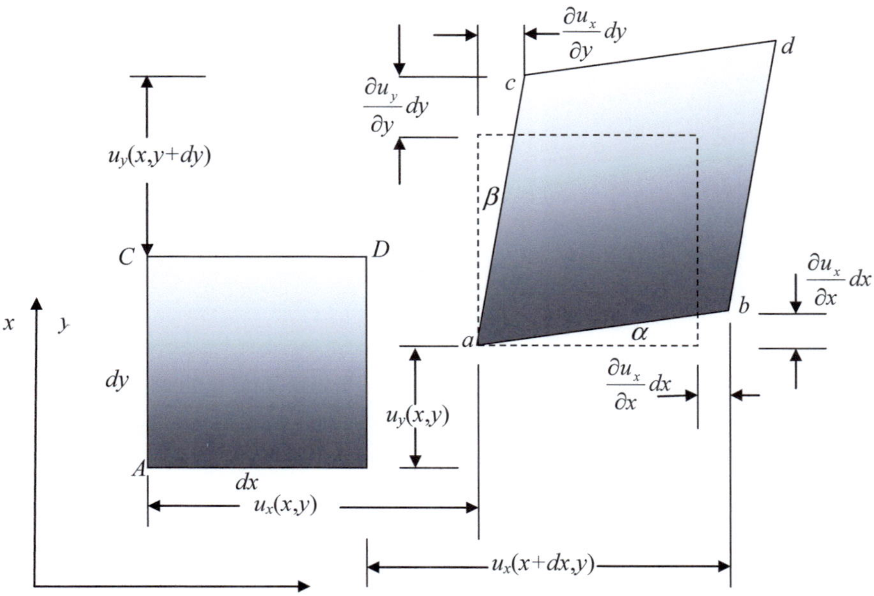

Fig. 2.26 2D strain

Now, consider two-dimensional deformation of an infinitesimal rectangular material element ABCD with dimensions $dx \times dy$, which after deformation, takes the form of a rhombus abcd in Fig. 2.26. Let the point A be displaced to a through $u_x(x,y)$ in x direction and $u_y(x,y)$ in y direction.

Variation of u_x in x direction is given by $\frac{\partial u_x}{\partial x}$, therefore the total variation over the distance dx is $\frac{\partial u_x}{\partial x}dx$. This gives us the direct strain ε_{xx}

$$\varepsilon_{xx} = \frac{\partial u_x}{\partial x} = u_{x,x} \tag{2.37a}$$

Similarly the variation of u_y in y direction is given by $\frac{\partial u_y}{\partial y}$, therefore the total variation over the distance dy is $\frac{\partial u_y}{\partial y}dy$. This gives us the direct strain ε_{yy}

$$\varepsilon_{yy} = \frac{\partial u_y}{\partial y} = u_{y,y} \tag{2.37b}$$

Next consider the point C whose displacement in x direction with AC rotating clockwise by infinitesimal angle β in the displaced position shown by dashed lines of the element $dx \times dy$ with A moving to a is given by the variation of u_x in y direction $\frac{\partial u_x}{\partial y}$. Total variation over dy is the shear strain ε_{xy}.

$$\beta \approx \tan \beta = \frac{\frac{\partial u_x}{\partial y} dy}{dy} = \frac{\partial u_x}{\partial y} = u_{x,y} \tag{2.37c}$$

Similarly the displacement of B in y direction with AB rotating counter-clockwise by infinitesimal angle α is given by variation of u_y in x direction $\frac{\partial u_y}{\partial x}$. Total variation over dx is the shear strain ε_{yx}.

$$\alpha \approx \tan \alpha = \frac{\frac{\partial u_y}{\partial x} dx}{dx} = \frac{\partial u_y}{\partial x} = u_{y,x} \tag{2.37d}$$

Summarizing the 2D strain can now be expressed as follows:
Direct strains:

$$\varepsilon_{xx} = \frac{\partial u_x}{\partial x} = u_{x,x} \quad \varepsilon_{yy} = \frac{\partial u_y}{\partial y} = u_{y,y}$$

Shear strains:

$$\varepsilon_{xy} = \frac{\partial u_x}{\partial y} = u_{x,y} \quad \varepsilon_{yx} = \frac{\partial u_y}{\partial x} = u_{y,x} \tag{2.37e}$$

Engineers prefer to use total shear strain $\gamma = \alpha + \beta$ which is twice of mathematical strains ε. While there is no difference in normal strains, the *engineering shear strains* are total strains which are twice that of mathematical shear strains.

Given the three displacements u_x, u_y and u_z, all the six strains in Eq. (2.37) are determinable whereas the inverse problem cannot be uniquely solved. The three displacements are to be obtained by solving the 6 partial differential equations therefore any given strain field cannot be associated with a unique displacement field. Also we have three stress equilibrium relations (2.4) and together with six strain-displacement relations above, we have 9 coupled partial differential equations with 15 independent parameters, viz., 3 displacements, 6 stresses and 6 strains. We need displacements, strains and stresses for a general three dimensional body under a given traction load. A discussion of Physics of deformable bodies considering stress and strain alone is still incomplete and we have to find additional 6 equations to be able to solve a structural problem.

Worked Example 2.7

Consider the case when all strains $\varepsilon_{ij} = u_{i,j} + u_{j,i} = 0$. Show that the integration of these equations leads to rigid body rotation.

Let u, v and w be the displacements; we first write the conditions that the normal strains vanish giving the following

$$\frac{\partial u}{\partial x} = 0 \Rightarrow u = u_1(y, z)$$
$$\frac{\partial v}{\partial y} = 0 \Rightarrow v = v_1(x, z)$$
$$\frac{\partial w}{\partial z} = 0 \Rightarrow w = w_1(x, y)$$

The rest of the strains are written using the above

$$\frac{\partial u}{\partial y} + \frac{\partial v}{\partial x} = 0 \Rightarrow \frac{\partial u_1}{\partial y} + \frac{\partial v_1}{\partial x} = 0 \tag{a}$$

$$\frac{\partial u}{\partial z} + \frac{\partial w}{\partial x} = 0 \Rightarrow \frac{\partial u_1}{\partial z} + \frac{\partial w_1}{\partial x} = 0 \tag{b}$$

$$\frac{\partial v}{\partial z} + \frac{\partial w}{\partial y} = 0 \Rightarrow \frac{\partial v_1}{\partial z} + \frac{\partial w_1}{\partial y} = 0 \tag{c}$$

Integrating (a) with respect to y and x gives the following relations

$$u_1 = -y\frac{\partial v_1}{\partial x} + f(z) \tag{a$'$}$$

$$v_1 = -x\frac{\partial u_1}{\partial y} + h(z) \tag{a$''$}$$

Similarly integrating (b) with respect to x and z gives the following relations

$$w_1 = -x\frac{\partial u_1}{\partial z} + j(y) \tag{b$'$}$$

$$u_1 = -z\frac{\partial w_1}{\partial x} + g(y) \tag{b$''$}$$

Finally integrating (c) with respect to z and y gives the following relations

$$v_1 = -z\frac{\partial w_1}{\partial y} + i(x) \tag{c$'$}$$

$$w_1 = -y\frac{\partial v_1}{\partial z} + k(x) \tag{c$''$}$$

For the normal strain $\frac{\partial u_1}{\partial x} = 0$ using (a″) and (b″) we get

$$\frac{\partial u_1}{\partial x} = 0 \Rightarrow \frac{\partial^2 v_1}{\partial x^2} = \frac{\partial^2 w_1}{\partial x^2} = 0$$
$$\therefore v_1 = A(z)x + B(z); \; \mathrm{w}_1 = C(y)x + D(y)$$

Similarly,

$$\frac{\partial v_1}{\partial y} = 0$$
$$\Rightarrow u_1 = E(z)y + F(z); \; \mathrm{w}_1 = G(x)y + H(x)$$
$$\frac{\partial w_1}{\partial z} = 0$$
$$\Rightarrow u_1 = I(y)z + J(y); \; \mathrm{v}_1 = K(x)z + L(x)$$

For the displacement u_1 we have from the above

$$u_1(y,z) = E(z)y + F(z) = I(y)z + J(y) \Rightarrow E(z)y - I(y)z = J(y) - F(z)$$

From the above we find $E\,(z) = E_1$ is a constant and $I(y) = I_1$ is another constant. Therefore

$$u_1(y,z) = E_1 y + I_1 z$$

Similarly, we arrive at

$$v_1(x,z) = K_1 z + A_1 x$$
$$w_1(x,y) = C_1 x + G_1 y$$

From (a) we have $\frac{\partial u_1}{\partial y} + \frac{\partial v_1}{\partial x} = E_1 + A_1 = 0 \Rightarrow E_1 = -A_1$

Similarly $C_1 = -I_1$ and $G_1 = -K_1$

We can combine the above results and arrive at the displacement field u, v, and w as functions of x, y, and z

$$u(x,y,z) = 0 \times x + E_1 y + I_1 z$$
$$v(x,y,z) = -E_1 x + 0 \times y + K_1 z$$
$$w(x,y,z) = -I_1 x - K_1 y + 0 \times z$$

The above is

$$\begin{Bmatrix} u \\ v \\ w \end{Bmatrix} = \begin{bmatrix} 0 & E_1 & I_1 \\ -E_1 & 0 & K_1 \\ -I_1 & -K_1 & 0 \end{bmatrix} \begin{Bmatrix} x \\ y \\ z \end{Bmatrix}$$

showing that there is a rigid body rotation.

2.4.2 Compatibility Conditions

Any given strain field cannot determine uniquely the three displacements; therefore, we determine the conditions required to make the strain field to be compatible to the displacement field by considering the strain displacement relations.

$$\varepsilon_{ij} = \frac{1}{2}\left(\frac{\partial u_i}{\partial x_j} + \frac{\partial u_j}{\partial x_i}\right) \tag{2.36}$$

We can first form the following derivatives

$$\begin{aligned} \varepsilon_{ij,kl} &= \frac{1}{2}\left(u_{i,jkl} + u_{j,ikl}\right) \\ \varepsilon_{kl,ij} &= \frac{1}{2}\left(u_{k,lij} + u_{l,kij}\right) \\ \varepsilon_{lj,ki} &= \frac{1}{2}\left(u_{l,jki} + u_{j,lki}\right) \\ \varepsilon_{ki,lj} &= \frac{1}{2}\left(u_{k,ilj} + u_{i,klj}\right) \end{aligned} \tag{2.38}$$

Adding the first two equations and then subtracting the last two equations, we can eliminate the displacements to give

$$\varepsilon_{ij,kl} + \varepsilon_{kl,ij} - \varepsilon_{lj,ki} - \varepsilon_{ki,lj} = 0 \tag{2.39}$$

It can be shown that there are 81 compatibility conditions from the above out of which six are independent, given by

$$\begin{aligned} &\frac{\partial^2 \varepsilon_{xx}}{\partial y^2} + \frac{\partial^2 \varepsilon_{yy}}{\partial x^2} = 2\frac{\partial^2 \varepsilon_{xy}}{\partial x \partial y} \quad \frac{\partial^2 \varepsilon_{xx}}{\partial y \partial z} = \frac{\partial}{\partial x}\left(-\frac{\partial \varepsilon_{yz}}{\partial x} + \frac{\partial \varepsilon_{zx}}{\partial y} + \frac{\partial \varepsilon_{xy}}{\partial z}\right) \\ &\frac{\partial^2 \varepsilon_{yy}}{\partial z^2} + \frac{\partial^2 \varepsilon_{zz}}{\partial y^2} = 2\frac{\partial^2 \varepsilon_{yz}}{\partial y \partial z} \quad \frac{\partial^2 \varepsilon_{yy}}{\partial z \partial x} = \frac{\partial}{\partial y}\left(\frac{\partial \varepsilon_{yz}}{\partial x} - \frac{\partial \varepsilon_{zx}}{\partial y} + \frac{\partial \varepsilon_{xy}}{\partial z}\right) \\ &\frac{\partial^2 \varepsilon_{zz}}{\partial x^2} + \frac{\partial^2 \varepsilon_{xx}}{\partial z^2} = 2\frac{\partial^2 \varepsilon_{zx}}{\partial z \partial x} \quad \frac{\partial^2 \varepsilon_{zz}}{\partial x \partial y} = \frac{\partial}{\partial z}\left(\frac{\partial \varepsilon_{yz}}{\partial x} + \frac{\partial \varepsilon_{zx}}{\partial y} - \frac{\partial \varepsilon_{xy}}{\partial z}\right) \end{aligned} \tag{2.40}$$

In two dimensional problems (e.g. plane strain), all z terms are set to zero. The compatibility equations reduce to just one

$$\frac{\partial^2 \varepsilon_{xx}}{\partial y^2} + \frac{\partial^2 \varepsilon_{yy}}{\partial x^2} = 2\frac{\partial^2 \varepsilon_{xy}}{\partial x \partial y} \tag{2.40a}$$

Let us now take stock of the equations we have. For general state of stress, we have three equations of equilibrium and six strain displacement relations, a total of nine equations. We also have six compatibility conditions that give us fifteen equations. The stresses are related to the applied tractions and are independent of strains. The strains are related to displacements and stresses do not enter into picture here.

Now the unknowns are the quantities that we want to determine to define the deformation problem; three displacements, six stresses and six strains that are fifteen quantities which are our unknowns. Summarizing

- Equilibrium Equations

$$\begin{aligned}
\frac{\partial \sigma_{xx}}{\partial x} + \frac{\partial \sigma_{yx}}{\partial y} + \frac{\partial \sigma_{zx}}{\partial z} + b_x = 0\\
\frac{\partial \sigma_{xy}}{\partial x} + \frac{\partial \sigma_{yy}}{\partial y} + \frac{\partial \sigma_{zy}}{\partial z} + b_y = 0\\
\frac{\partial \sigma_{xz}}{\partial x} + \frac{\partial \sigma_{yz}}{\partial y} + \frac{\partial \sigma_{zz}}{\partial z} + b_z = 0
\end{aligned} \tag{2.4}$$

- Strain Displacement Relations

$$\begin{aligned}
\varepsilon_{xx} &= \frac{\partial u}{\partial x} & \varepsilon_{yz} &= \frac{1}{2}\left(\frac{\partial v}{\partial z} + \frac{\partial w}{\partial y}\right) = \varepsilon_{zy}\\
\varepsilon_{yy} &= \frac{\partial v}{\partial y} & \varepsilon_{zx} &= \frac{1}{2}\left(\frac{\partial w}{\partial x} + \frac{\partial u}{\partial z}\right) = \varepsilon_{xz}\\
\varepsilon_{zz} &= \frac{\partial w}{\partial z} & \varepsilon_{xy} &= \frac{1}{2}\left(\frac{\partial u}{\partial y} + \frac{\partial v}{\partial x}\right) = \varepsilon_{yx}
\end{aligned} \tag{2.37}$$

- Compatibility Conditions

$$\begin{aligned}
\frac{\partial^2 \varepsilon_{xx}}{\partial y^2} + \frac{\partial^2 \varepsilon_{yy}}{\partial x^2} = 2\frac{\partial^2 \varepsilon_{xy}}{\partial x \partial y} \quad & \frac{\partial^2 \varepsilon_{xx}}{\partial y \partial z} = \frac{\partial}{\partial x}\left(-\frac{\partial \varepsilon_{yz}}{\partial x} + \frac{\partial \varepsilon_{zx}}{\partial y} + \frac{\partial \varepsilon_{xy}}{\partial z}\right)\\
\frac{\partial^2 \varepsilon_{yy}}{\partial z^2} + \frac{\partial^2 \varepsilon_{zz}}{\partial y^2} = 2\frac{\partial^2 \varepsilon_{yz}}{\partial y \partial z} \quad & \frac{\partial^2 \varepsilon_{yy}}{\partial z \partial x} = \frac{\partial}{\partial y}\left(\frac{\partial \varepsilon_{yz}}{\partial x} - \frac{\partial \varepsilon_{zx}}{\partial y} + \frac{\partial \varepsilon_{xy}}{\partial z}\right)\\
\frac{\partial^2 \varepsilon_{zz}}{\partial x^2} + \frac{\partial^2 \varepsilon_{xx}}{\partial z^2} = 2\frac{\partial^2 \varepsilon_{zx}}{\partial z \partial x} \quad & \frac{\partial^2 \varepsilon_{zz}}{\partial x \partial y} = \frac{\partial}{\partial z}\left(\frac{\partial \varepsilon_{yz}}{\partial x} + \frac{\partial \varepsilon_{zx}}{\partial y} - \frac{\partial \varepsilon_{xy}}{\partial z}\right)
\end{aligned} \tag{2.40}$$

It sounds good that we have 15 equations and 15 unknowns and we can buckle up and find ways of solving them. When we try and do that we face another problem, the first three equations in (2.4) are totally unconnected with the rest of 12 equations and the first three cannot be solved independently. Something else is missing, the solid and its characteristics. If the solid is not playing a role, we will have same solutions for all materials, e.g., steel and rubber foam. We have to bring this material property into the scene to find a deformation under load.

2.5 Stress-Strain Relations

Robert Hooke, a contemporary of Newton in 1660 discovered the law of elasticity which bears his name and which describes the linear variation of tension with extension in an elastic spring. We find from tests that stresses and strains are related amongst themselves through the material characteristics. That's how different materials find applications under different conditions.

2.5.1 Tension Test

Robert Hooke used experimental apparatus like that shown in Fig. 2.27 to measure the extension of a spring as a function of the imposed load and thereby formulated his famous law; "As the extension, so is the force." In modern parlance, this is $\sigma = E\varepsilon$; Stress is proportional to Strain with the constant of proportionality given by the Young's modulus, E.

Uniaxial tension test is commonly used to determine the stress-strain behavior of isotropic materials (materials having identical properties in any direction). A typical modern tension testing machine is shown in Fig. 2.28a. As these tests are meant for identifying the material behavior, one has to follow standards that are established by societies such as American Society for Testing Materials. In particular testing specimens are standardized such as the ones shown in Fig. 2.28b.

A standard tension test specimen typically has an original diameter $d_0 = 2.5, 6.25$ or 12.5 mm and common gauge lengths l_0 adopted are 10, 25 or 50 mm respectively. The specimen is mounted in a tensile test machine and slowly (gradually) loaded in tension. The load is static and while increasing the load, care is to be taken so that no reversals of stress take place. Such a test is called static or monotonically loaded test. At different loads thus applied, the load and the gauge lengths are observed. The results are plotted on stress-strain diagram. Both the stress and strain are based on the original diameter and gauge length—these parameters change during the test and these changes can be neglected for elastic deformations.

The strain is determined using the original gauge length l_0 and instantaneous length l_i that is the nominal or engineering strain as shown in Fig. 2.29. During the test the cross-section of the specimen decreases, in elastic region this is negligible. Nominal

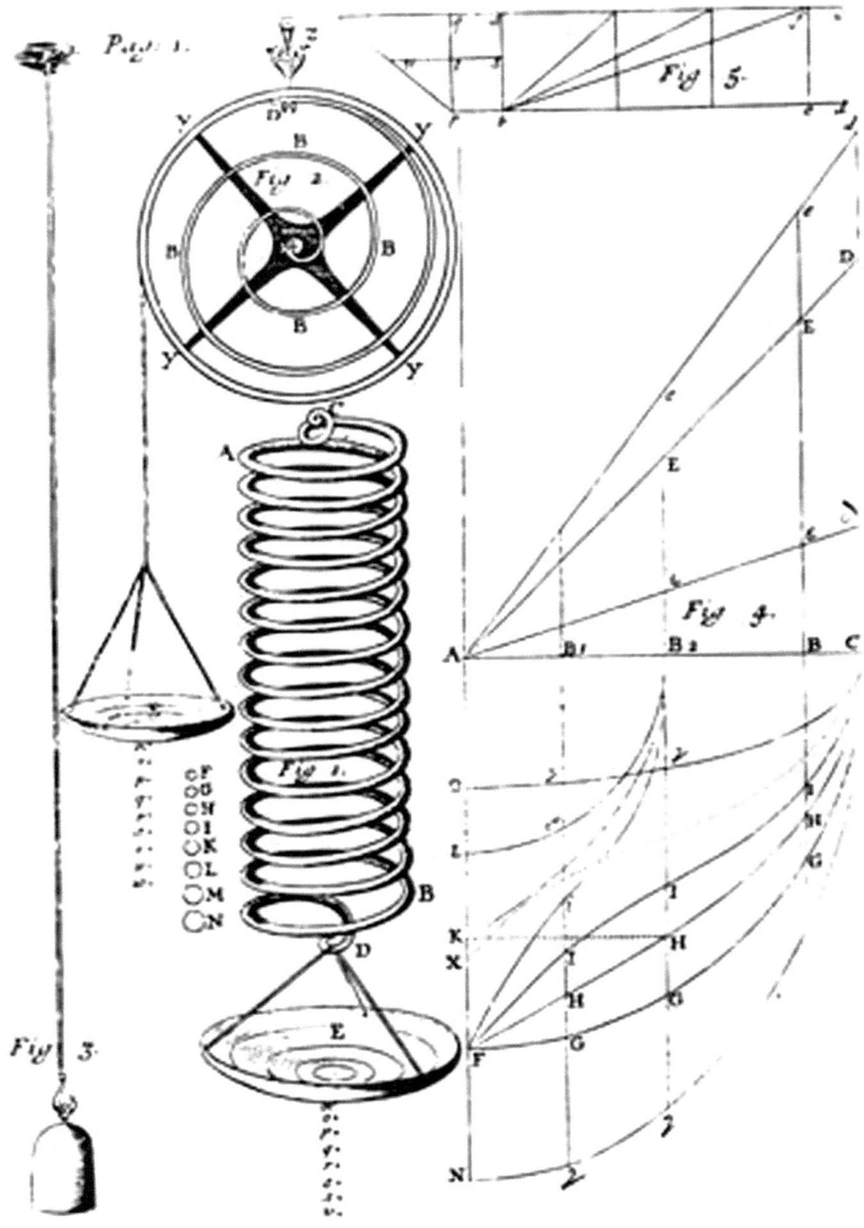

Fig. 2.27 Robert Hooke's apparatus to determine elasticity of a spring

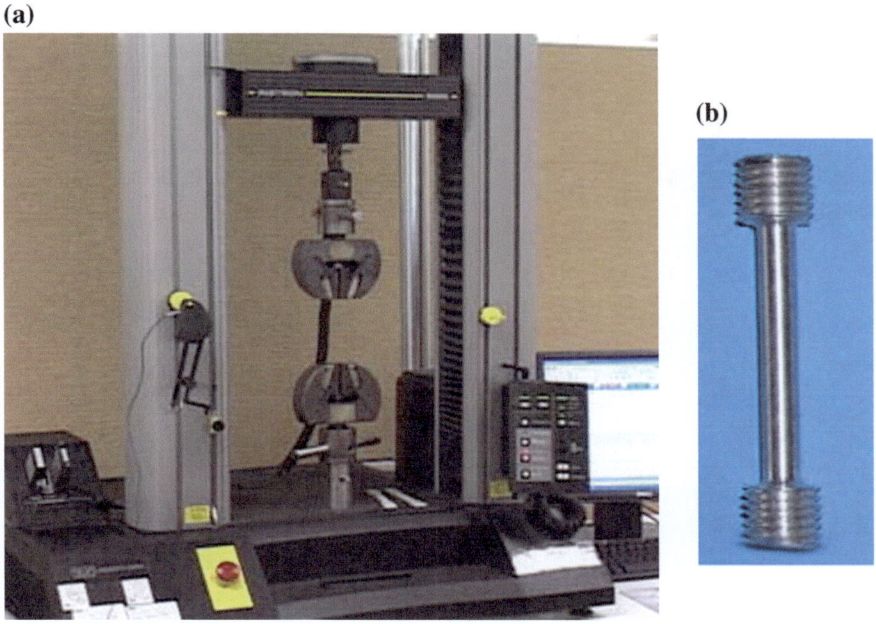

Fig. 2.28 Tension testing machine with a specimen

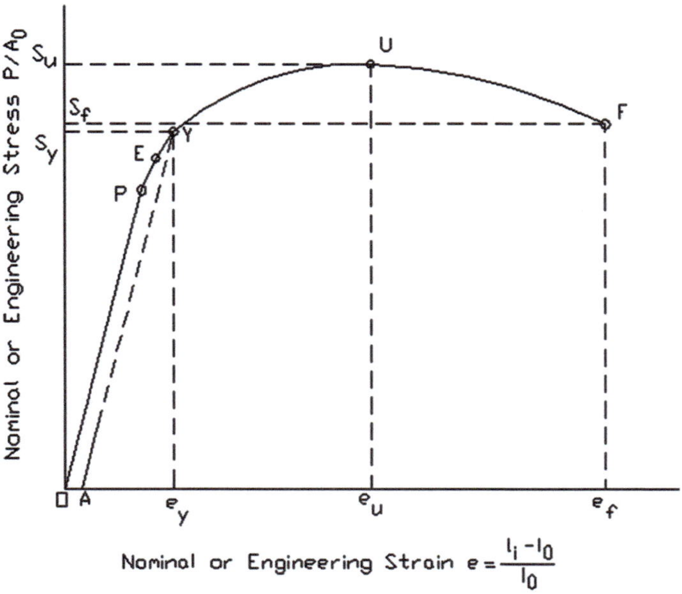

Fig. 2.29 Stress strain diagram of an isotropic material

or engineering stress S is defined by P/A_0 ignoring the change in cross-section. We generally adopt structures to be operating in elastic regime; in plastic region, the material suffers large strains and therefore they are not safe. Therefore the stress-strain diagram is usually considered with nominal stress and strain for the study.

As we gradually load the specimen (avoiding any reversals), we observe Hooke's law valid till we reach point P. In this region nominal stress is proportional to nominal strain, i.e.,

$$S = Ee \tag{2.41}$$

P is called *Proportionality Limit*. The slope of the line OP gives *elastic modulus*, also called *Young's modulus* $E = \frac{S}{e}$. Thomas Young is a famous elastician of 18th century who described the characterization of elasticity in 1807 that came to be known as Young's modulus. Because we made approximation in defining stress and strain by ignoring the change in cross-sectional dimension, the stress and strain in the elastic region are denoted S and e to distinguish them from true values of stress and strain denoted by σ and ε. After the proportionality is reached, the linear law of Hooke is not valid. With a slight increase in the load, we reach the point E which is the *Elastic Limit*. In the region P to E the linearity is lost; however, if we unload the specimen, we revert on the same line and reach the original condition when the load is completely removed. This is Elastic region, i.e., there is no permanent set up to this point upon removal of load.

When we increase the load further from the elastic limit, the material begins to yield under the load rather than resisting the load as in elastic region. The strain begins to increase rapidly and the material is termed to flow as a plastic material. The region beyond elastic limit is not well definable. i.e., we cannot find exactly when the yielding begins. Some materials relax at E before showing some resistance, though much less than in elastic region. It is a standard practice to define yield condition based on the permanent set on gauge length. When we reach 0.2 to 0.5 % permanent set on gauge length, the material is said to yield denoted by point Y on Fig. 2.29. Yield is always specified together with permanent set. The stress at this point is denoted S_y as shown in Fig. 2.29.

Upon increasing the load, the strain rapidly increases as if it is flowing compared to elastic region when the strain is very low. The material goes into plastic region. Even in the plastic region, it offers resistance, though much less than elastic region. It is as if the material gets tired and is in no mood to resist. Also the definition of elasticity is lost and the strain based on the original gauge length is erroneous, we should apply the true strain condition. The true strain is defined by

$$\begin{aligned} \varepsilon &= \int\limits_{l_0}^{l_i} \frac{dl}{l} = \ln \frac{l_i}{l_0} \\ &= \ln(1+e) \end{aligned} \tag{2.42}$$

in place of engineering strain defined in Eq. (2.32). Since the area of cross-section decreases in this plastic region from the original area A, we also have to consider true stress P/A_i in place of engineering stress P/A. Since no structure experienced plastic condition until recent times, this condition was ignored and the stress strain diagram continued to be defined assuming that the engineering stress and strain are valid beyond yield.

Point U in Fig. 2.29 denotes the Ultimate (Engineering) Stress S_u, called *Ultimate Tensile Strength* representing the maximum engineering stress reached in this test. Even beyond the point U, we continue to use the original cross-section area A, the engineering stress decreases with strain rapidly increasing, i.e., the test specimen elongating rapidly till fracture occurs at F. Stresses used here are not true stresses and *Necking* takes place in this region before ultimate failure, see Fig. 2.30. The stress at F is denoted by *Fracture Stress* S_f. The true value of stress here is more than Ultimate stress and fracture is typically cup-and-cone fracture in shear at an angle 45°.

We can plot the same experimental data in true stress—true strain axes rather than engineering stress—engineering strain axes; then we get Fig. 2.31. In the elastic part, there will be practically no difference as the true stress and strain are not different in this region. Beyond yield, the material flows, that is the true strain increases rapidly. The erroneous interpretation in engineering stress and strain plot on the fracture stress being lower than that of the ultimate stress is corrected in Fig. 2.31.

In Fig. 2.31, the corresponding points of ultimate and fracture points in engineering stress-strain plot are also shown. The plastic region is described by $\sigma = K\varepsilon^n$ where K is strength coefficient and n is strength exponent. In a log-log plot, the plastic region is a straight line that allows us to determine these material properties K and n. The analysis of stress and strain in the plastic region is beyond the scope for the present and we restrict to elasticity part only in what follows.

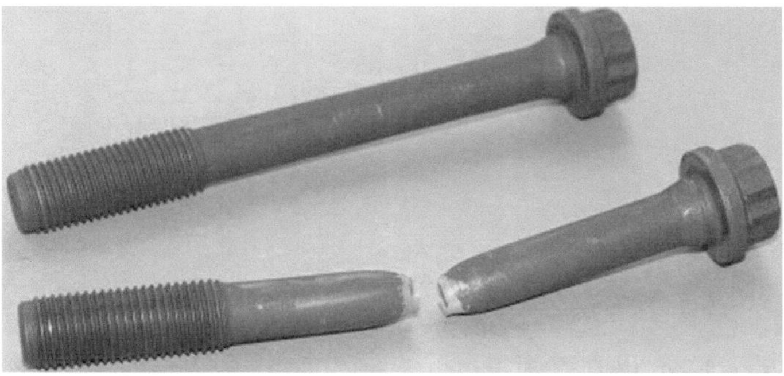

Fig. 2.30 Cup and cone fracture

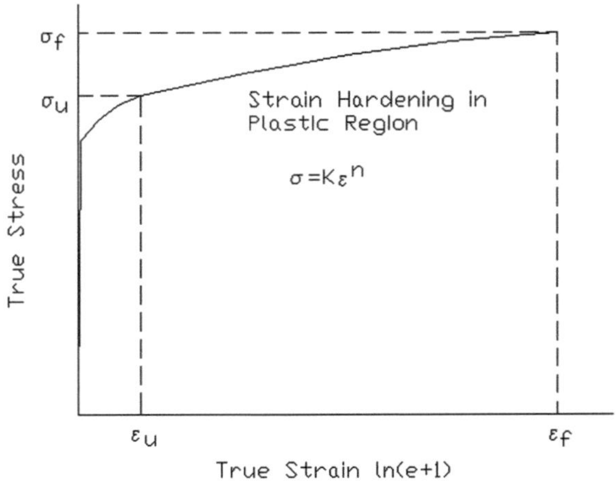

Fig. 2.31 True stress-strain diagram

2.5.2 *Shear Modulus* G *(Rigidity Modulus)*

The behavior of isotropic materials like mild steel is similar to tension load. The slope of the elastic line in shear is called *shear modulus* or *rigidity modulus* denoted by G which is the ratio of shear stress to shear strain in the elastic region.

2.5.3 *Compression or Bulk Modulus* K

The compressive stress-strain curve is similar to the tensile stress-strain curve up to the yield point. Thereafter, the progressively increasing specimen cross section causes the compressive stress-strain curve to diverge from the tensile curve. The ratio of compressive stress to compressive strain in the elastic region is called *Bulk modulus* denoted by K.

2.5.4 *Poisson's Ratio* ν

In a tension test, we noted that the cross-sectional area changes and this change is related to the elongation by Poisson's ratio ν which is the ratio of strain in a transverse direction to that in the longitudinal direction. Siméon Denis Poisson is an 18th century French mathematician.

2.5.5 General Symmetric Linear Stress-Strain Behavior

In general a body is subjected to six components of stress, therefore we can think of most general symmetric linear stress-strain relation to be

$$\begin{Bmatrix} \sigma_{xx} \\ \sigma_{yy} \\ \sigma_{zz} \\ \sigma_{xy} \\ \sigma_{yz} \\ \sigma_{zx} \end{Bmatrix} = \begin{bmatrix} C_{xxxx} & C_{xxyy} & C_{xxzz} & C_{xxxy} & C_{xxyz} & C_{xxzx} \\ & C_{yyyy} & C_{yyzz} & C_{yyxy} & C_{yyyz} & C_{yyzx} \\ & & C_{zzzz} & C_{zzxy} & C_{zzyz} & C_{zzzx} \\ & & & C_{xyxy} & C_{xyyz} & C_{xyzx} \\ & & & & C_{yzyz} & C_{yzzx} \\ & & & & & C_{zxzx} \end{bmatrix} \begin{Bmatrix} \varepsilon_{xx} \\ \varepsilon_{yy} \\ \varepsilon_{zz} \\ \varepsilon_{xy} \\ \varepsilon_{yz} \\ \varepsilon_{zx} \end{Bmatrix} \tag{2.43}$$

Here we assume that every stress is dependent on all strains and are related linearly giving rise to 21 coefficients or constants of the material. Most of the materials behave linearly as above before the material yields and then we have linear elasticity; otherwise they are plastic in nature. We will keep them out in this discussion.

Equation (2.43) also assumes that there is complete absence of isotropy, i.e., the properties are different in different directions. Though many materials are anisotropic, fortunately a good number of materials are isotropic that makes our life easier. Equation (2.43) is simplified to

$$\begin{Bmatrix} \sigma_{xx} \\ \sigma_{yy} \\ \sigma_{zz} \\ \sigma_{xy} \\ \sigma_{yz} \\ \sigma_{zx} \end{Bmatrix} = \begin{bmatrix} \lambda+2G & \lambda & \lambda & & & \\ \lambda & \lambda+2G & \lambda & & & \\ \lambda & \lambda & \lambda+2G & & & \\ & & & 2G & & \\ & & & & 2G & \\ & & & & & 2G \end{bmatrix} \begin{Bmatrix} \varepsilon_{xx} \\ \varepsilon_{yy} \\ \varepsilon_{zz} \\ \varepsilon_{xy} \\ \varepsilon_{yz} \\ \varepsilon_{zx} \end{Bmatrix} \tag{2.44}$$

λ is called Lamé's first parameter and together with G they are called Lamé's parameters. They are related to Young's modulus E and Poisson's ratio v. Gabriel Léon Jean Baptiste Lamé is a 19th century French mathematician. Lame's constants are

$$\begin{aligned} &G = \frac{E}{2(1+v)}; \quad \lambda = \frac{Ev}{(1+v)(1-2v)} \\ &\text{or} \\ &E = \frac{G(3\lambda+2G)}{\lambda+G}; \quad v = \frac{\lambda}{2(\lambda+G)} \end{aligned} \tag{2.45}$$

Equations (2.44) can be written in a compact form relating stresses with strains through

$$\sigma_{ij} = \lambda\delta_{ij}\varepsilon_{ii} + 2G\varepsilon_{ij} \tag{2.46}$$

In a simple tension test where the change in the cross-sectional area is ignored, we set $\nu \to 0$

$$\begin{aligned} &\lambda + 2G = E \\ &\text{and} \\ &\sigma = E\varepsilon \end{aligned} \tag{2.47}$$

This defines Young's modulus E. We can also derive strains in terms of stresses through

$$\varepsilon_{ij} = \frac{1}{2G}\sigma_{ij} - \left[\frac{\lambda + G}{2G(3\lambda + G)}\sigma_{ii}\right]\delta_{ij} \tag{2.48}$$

Now we can convert Eq. (2.4) from stresses to strains and together with strain displacement relations and compatibility conditions we can solve the problem. Before we discuss the method of solving the structural problem, we need the strain energy of the internal stress and strain condition to balance the external work from the applied traction loads, as given in the next section.

Worked Example 2.8

Using the stress-strain law in (2.46), consider the uniaxial tension test in z direction, i.e., $\sigma_{zz} \neq 0, \sigma_{xx} = 0, \sigma_{yy} = 0$ and relate the strains ε_{xx} and ε_{yy} to ε_{zz} and σ_{zz} to ε_{zz}. Compare the analytical results with experimental data $\varepsilon_{xx} = \varepsilon_{yy} = -\nu\varepsilon_{zz}$ and $\sigma_{zz} = E\varepsilon_{zz}$ and obtain the result in (2.45).

Using (2.46) $\sigma_{ij} = \lambda\delta_{ij}\varepsilon_{ii} + 2G\varepsilon_{ij}$, we have for this case

$$\sigma_{zz} = \lambda\varepsilon_{xx} + \lambda\varepsilon_{yy} + (\lambda + 2G)\varepsilon_{zz}$$

and

$$\begin{aligned} \sigma_{xx} &= 0 = (\lambda + 2G)\varepsilon_{xx} + \lambda\varepsilon_{yy} + \lambda\varepsilon_{zz} \\ \sigma_{yy} &= 0 = \lambda\varepsilon_{xx} + (\lambda + 2G)\varepsilon_{yy} + \lambda\varepsilon_{zz} \end{aligned}$$

Solving the above last two equations for ε_{xx}, ε_{yy}

$$\varepsilon_{xx} = \varepsilon_{yy} = -\frac{\lambda}{2(\lambda + G)}\varepsilon_{zz}$$

then

$$\sigma_{zz} = \left[(\lambda + 2G) - \frac{2\lambda^2}{2(\lambda + G)} \right] \varepsilon_{zz} = \frac{G(2G + 3\lambda)}{\lambda + G} \varepsilon_{zz}$$

From one dimensional test data $\sigma_{zz} = E\varepsilon_{zz} \therefore E = \frac{G(2G+3\lambda)}{\lambda+G}$ and $\varepsilon_{xx} = \varepsilon_{yy} = -\nu\varepsilon_{zz} \because \nu = \frac{\lambda}{2(\lambda+G)} \Rightarrow \lambda = \frac{2\nu G}{(1-2\nu)}$. This gives $G = \frac{E}{2(1+\nu)}$ and $\lambda = \frac{E\nu}{(1+\nu)(1-2\nu)}$.

2.6 Strain Energy and Work

Internally when the body deforms we have a stress field in the entire body. At any point we have also a strain field that represents the deformation; there is a work done by the strain field on the stress field. This internal energy over the whole volume of the body balances the external work done by the applied traction loads.

This strain energy can be calculated by considering the infinitesimal rectangular parallelepiped under the normal and shear stresses as in Fig. 2.9. Considering x direction, the displacements of the left and right faces are given by u_x and $u_x + \left(\frac{\partial u_x}{\partial x}\right)dx$.

The mechanical work done by direct stress σ_{xx} during deformation in x direction is

$$dw_x = -\sigma_{xx} du_x dydz + \left(\sigma_{xx} + \frac{\partial \sigma_{xx}}{\partial x} dx \right) d\left(u_x + \frac{\partial u_x}{\partial x} dx \right) dydz \tag{2.49}$$

Simplifying and neglecting higher order terms and considering equilibrium relations

$$\begin{aligned} dw_x &= \left\{ \sigma_{xx} d\left(\frac{\partial u_x}{\partial x} \right) + \left(\frac{\partial \sigma_{xx}}{\partial x} \right) du_x \right\} dxdydz \\ &= \left\{ \sigma_{xx} d\left(\frac{\partial u_x}{\partial x} \right) \right\} dxdydz = \sigma_{xx} d\varepsilon_{xx} dv \end{aligned} \tag{2.50}$$

Considering similarly y and z, the increment in mechanical work from direct stresses is

$$dw_d = \left(\sigma_{xx} d\varepsilon_{xx} + \sigma_{yy} d\varepsilon_{yy} + \sigma_{zz} d\varepsilon_{zz} \right) dv \tag{2.55}$$

Similarly the work from shear stresses is

$$dw_s = \left(2\sigma_{xy} d\varepsilon_{xy} + 2\sigma_{xz} d\varepsilon_{xz} + \sigma_{yz} d\varepsilon_{yz} \right) dv \tag{2.56}$$

Summing up Eqs. (2.55) and (2.56), the total mechanical work of the stresses is

$$\begin{aligned} dw &= \left(\sigma_{xx}d\varepsilon_{xx} + \sigma_{yy}d\varepsilon_{yy} + \sigma_{zz}d\varepsilon_{zz} + 2\sigma_{xy}d\varepsilon_{xy} + 2\sigma_{xz}d\varepsilon_{xz} + \sigma_{yz}d\varepsilon_{yz}\right)dv \\ &= \sigma_{ij}d\varepsilon_{ij}dv \end{aligned} \tag{2.57}$$

$$\therefore w = \int_0^{\varepsilon_{ij}} \sigma_{ij}d\varepsilon_{ij} = U_0 \tag{2.58}$$

The work done by the strain field on the stress field at point *P* is the strain energy density U_0 given above. For U_0 to be a Point function the integral must be independent of the path, $\sigma_{ij}d\varepsilon_{ij}$ must be a perfect differential. i.e.,

$$dU_0 = \sigma_{ij}d\varepsilon_{ij} \tag{2.59}$$

This implies

$$\frac{dU_0}{d\varepsilon_{ij}} = \sigma_{ij} \tag{2.60}$$

Every loaded structure has strain energy given by

$$U = \iiint\limits_v U_0 dv = \iiint\limits_v v\left[\int_0^{\varepsilon_{ij}} \sigma_{ij}d\varepsilon_{ij}\right]dv = \frac{1}{2}\iiint\limits_v v\sigma_{ij}\varepsilon_{ij}dv \tag{2.61}$$

The strain energy density function in general can be written as

$$U_0 = \frac{1}{2}\left(\sigma_{xx}\varepsilon_{xx} + \sigma_{yy}\varepsilon_{yy} + \sigma_{zz}\varepsilon_{zz} + \sigma_{yz}\varepsilon_{yz} + \sigma_{zx}\varepsilon_{zx} + \sigma_{xy}\varepsilon_{xy}\right) \tag{2.62}$$

The strain energy density above can be further simplified by considering a coordinate system that is parallel to the principal stress directions. In this coordinate system, no shear components exist.

Therefore

$$U_0 = \frac{1}{2}(\sigma_1\varepsilon_1 + \sigma_2\varepsilon_2 + \sigma_3\varepsilon_3) \tag{2.63}$$

Writing the strains in principal coordinates in terms of the three principal stresses and the definition of Poisson's ratio, we have

$$\begin{aligned}
\varepsilon_1 &= \frac{1}{E}(\sigma_1 - \nu\sigma_2 - \nu\sigma_3) \\
\varepsilon_2 &= \frac{1}{E}(\sigma_2 - \nu\sigma_1 - \nu\sigma_3) \\
\varepsilon_3 &= \frac{1}{E}(\sigma_3 - \nu\sigma_1 - \nu\sigma_2)
\end{aligned} \tag{2.64}$$

Equation (2.63) for strain energy density can now be expressed in terms of principal stresses alone as

$$U_0 = \frac{1}{2E}\left[\sigma_1^2 + \sigma_2^2 + \sigma_3^2 - 2\nu(\sigma_1\sigma_2 + \sigma_2\sigma_3 + \sigma_1\sigma_3)\right] \tag{2.65}$$

Worked Example 2.9

For plane stress condition of an isotropic elastic body subjected to a temperature change ΔT the stress strain relations are given by

$$\begin{aligned}
\sigma_{xx} &= \frac{E}{1-\nu^2}\left(\varepsilon_{xx} + \nu\varepsilon_{yy}\right) - \frac{E}{1+\nu}\alpha\Delta T \\
\sigma_{yy} &= \frac{E}{1-\nu^2}\left(\nu\varepsilon_{xx} + \varepsilon_{yy}\right) - \frac{E}{1+\nu}\alpha\Delta T \\
\sigma_{xy} &= 2G\varepsilon_{xy}
\end{aligned}$$

where α is coefficient of thermal expansion. Determine the strain energy density U_0.

$$\begin{aligned}
U_0 &= \int_0^{\varepsilon_{xx}} \sigma_{xx} d\varepsilon_{xx} + 2\int_0^{\varepsilon_{xy}} \sigma_{xy} d\varepsilon_{xy} + \int_0^{\varepsilon_{yy}} \sigma_{yy} d\varepsilon_{yy} \\
&= \frac{E}{2(1-\nu^2)}\left(\varepsilon_{xx}^2 + 2\nu\varepsilon_{xx}\varepsilon_{yy} + \varepsilon_{yy}^2\right) + 2G\varepsilon_{xy}^2 - \frac{E\alpha\Delta T}{1+\nu}\left(\varepsilon_{xx} + \varepsilon_{yy}\right)
\end{aligned}$$

2.7 Von Mises Stress

Most of the material testing is done in uniaxial tension to determine the material properties, see Sect. 2.5. These material properties are used while designing a machine member. While we can use the general state of stress at a point and the planes on which principal or maximum shear stresses act, judiciously in a design it will be preferable to have a controlling design stress from analysis in a general 3D state. To facilitate this, several failure criteria are used, most common one being distortion or octahedral shear theories that lead to von Mises–Hencky relation.

The strain energy density in (2.65) can be thought of as consisting of two components: one due to dilation or change in volume and the other due to distortion or change in shape. The former is called dilatational strain energy and the latter distortional energy. Many experiments have shown that ductile materials can be hydrostatically stressed to levels beyond their ultimate strength in compression without failure. This is because the hydrostatic state of stress reduces the volume of the specimen without changing its shape.

The distortion energy theory states "*A machine member under multiaxial state of stress fails when the distortion energy per unit volume becomes equal to or exceeds the distortion energy per unit volume at the time of failure in a simple uniaxial stress test using a specimen of the same material*".

Decomposition of Strain Energy:

The strain energy density at a point in a solid can be divided into two parts: dilatational strain energy density U_h that is due to change in volume, and distortional strain energy density, U_d, that is responsible for change in shape. In order to compute these components, we divide the stress matrix also into similar components, dilatational stress matrix and deviatoric stress matrix. For convenience we will refer the stresses to the principal stress coordinates. Then

$$\begin{bmatrix} \sigma_1 & & \\ & \sigma_2 & \\ & & \sigma_3 \end{bmatrix} = \begin{bmatrix} \sigma_h & & \\ & \sigma_h & \\ & & \sigma_h \end{bmatrix} + \begin{bmatrix} \sigma_{1d} & & \\ & \sigma_{2d} & \\ & & \sigma_{3d} \end{bmatrix}$$

The dilatational stress component σ_h which is also called the volumetric stress is given by

$$\sigma_h = \frac{1}{3}(\sigma_1 + \sigma_2 + \sigma_3)$$

Substituting the above in (2.65), the dilatational part of the strain energy is

$$\begin{aligned} U_h &= \frac{1}{2E}\left[\sigma_h^2 + \sigma_h^2 + \sigma_h^2 - 2v(\sigma_h\sigma_h + \sigma_h\sigma_h + \sigma_h\sigma_h)\right] \\ &= \frac{3}{2}\frac{(1-2v)}{E}\sigma_h^2 \\ &= \frac{3}{2}\frac{(1-2v)}{E}\left[\frac{1}{3}(\sigma_1 + \sigma_2 + \sigma_3)\right]^2 \\ &= \frac{(1-2v)}{6E}\left[\sigma_1^2 + \sigma_2^2 + \sigma_3^2 + 2(\sigma_1\sigma_2 + \sigma_2\sigma_3 + \sigma_1\sigma_3)\right] \end{aligned}$$

The distortion part of the energy is therefore

$$\begin{aligned}U_d &= U_0 - U_h \\ &= \frac{1}{2E}\left[\sigma_1^2+\sigma_2^2+\sigma_3^2-2v(\sigma_1\sigma_2+\sigma_2\sigma_3+\sigma_1\sigma_3)\right] \\ &\quad -\frac{(1-2v)}{6E}\left[\sigma_1^2+\sigma_2^2+\sigma_3^2+2(\sigma_1\sigma_2+\sigma_2\sigma_3+\sigma_1\sigma_3)\right] \\ &= \frac{1+v}{3E}\left[\sigma_1^2+\sigma_2^2+\sigma_3^2-(\sigma_1\sigma_2+\sigma_2\sigma_3+\sigma_1\sigma_3)\right] \\ &= \frac{1+v}{3E}\left[\frac{(\sigma_1-\sigma_2)^2+(\sigma_2-\sigma_3)^2+(\sigma_3-\sigma_1)^2}{2}\right]\end{aligned}$$

Writing $U_d = \frac{1+v}{3E}\sigma_v^2$, we obtain von Mises stress from principal stresses, and this stress usually controls the design.

$$\sigma_v = \sqrt{\frac{1}{2}\left[(\sigma_1-\sigma_2)^2+(\sigma_2-\sigma_3)^2+(\sigma_3-\sigma_1)^2\right]} \tag{2.66}$$

This theory is usually adopted for all ductile materials. Note that the above stress is always positive, whether the principal stresses are compressive or tensile in nature.

Worked Example 2.10

In continuation to worked example 2.4, determine the equivalent stress.

$$\sigma_l = \frac{pr}{2t} = 69.33\,\text{MPa} \quad \sigma_h = \frac{pr}{t} = 138.667\,\text{MPa}$$

$$\sigma_{\text{eqv}} = \sqrt{(\sigma_l)^2+(\sigma_h)^2-\sigma_l\times\sigma_l} = 120.08\ \text{MPa}$$

2. Exercises

2.1. Discuss what external tractions are in a general three dimensional solid or structure. Take a simple example of an overhead cantilevered crane carrying a load; explain the tractions on the cantilever.
2.2. What are internal tractions? Discuss the importance of a datum on which the internal tractions act. Also explain the connection between the internal tractions that lead to stresses at a point on the chosen datum.
2.3. Take the example of the connecting rod in a scooter engine; isolate it from the piston (piston pin) and crank shaft (crankpin) and discuss how the reaction forces here can be considered as external tractions.

2.4. Discuss the general state of stress at a point on a datum plane in terms of a coordinate system suitably chosen. Use a second order tensor to describe the stress state. Show that there are three direct stresses (normal to the plane) and six shear (in-plane) stresses to describe the general state of stress.

2.5. Consider the state of general stress discussed in problem 4 on an infinitesimally small element with its sides parallel to the coordinate system chosen. Derive from fundamental principles the state of equilibrium in any one direction, say x considering the relevant direct and shear stresses varying across the infinitesimally small distances of the element. Assume that the body forces can be represented by a vector at the point under consideration and that it can be resolved into components along the directions of the coordinate system chosen.

2.6. Extend the state equation derived in problem 5 above and obtain the three equations of equilibrium written for the arbitrarily chosen the frame. Comment on the number of equations, three in this case, and the number of unknowns, (six or nine?), in the stress tensor.

2.7. Consider the connecting rod in Fig. 2.5 of this chapter, at one crank angle; this rod can be simplified to be carrying 1000 N force. Midway the length of the rod, the cross-sectional area A is 8 cm^2, determine the average stress in this cross-section. Comment on this value of average stress as compared to the stress tensor at a point, keeping in view the question 6 above.

2.8. The stress components at a point in a structure are given as $\begin{bmatrix} 7 & 1.5 & 0 \\ 1.5 & -40 & -3 \\ 0 & -3 & 0 \end{bmatrix}$ MPa. Find the components of the surface traction vector for an interface having a normal vector $\vec{v} = 0.10\vec{i} + 0.3\vec{j} + 0.8\vec{k}$.

2.9. The state of stress at a point in a structure is given to be $\begin{bmatrix} 20 & 10 & 0 \\ 10 & 0 & 0 \\ 0 & 0 & 50 \end{bmatrix}$ MPa. Rotate the current axes system by 60° about z axis and rewrite the stress tensor in the modified coordinate system.

2.10. Given σ_x = 80 MPa, σ_y = −50 MPa and σ_{xy} = 30 MPa, determine the principal stresses, their directions and maximum shear stress and the planes on which it acts. What is the normal stress on the maximum shear stress planes?

2.11. In the Eq. (2.4) of the text, we find that there are six unknowns with only three equations. Discuss the missing physics in this formulation and how to overcome this. Despite the apparent difficulty mentioned, demonstrate the usefulness of these equations, particularly from the transformation of coordinates of a given problem.

Chapter 3
Euler-Lagrange Equations

Every problem in Physics is set up through a set of partial differential equations subjected to constraints. Usually there is no exact solution excepting very rare cases. In general we have to find an approximate way of obtaining the desired solution. The most general approach to this attempt is through Variational Calculus which was invented by Isaac Newton in 1697. Physical phenomena are expressed as Functionals, that is Function of Functions, e.g., $F(x, y, y')$; here y is a function of x, where x is an independent parameter like the distance measured along a bar and y could be the deflection at any distance x. y' is first derivative of y with respect to x.

Consider an integral containing a functional of the form

$$I = \int_{x_1}^{x_2} F(x, y, y')dx \tag{3.1}$$

The value of I between (x_1, y_1) and (x_2, y_2) will depend on the path $y(x)$ chosen, see Fig. 3.1. We assume the existence of a path $y(x)$ that will have a property of extremizing I with respect to neighboring paths called varied paths denoted collectively by $\widehat{y}(x)$. Let this be written as $\widehat{y}(x) = y(x) + \varepsilon\eta(x)$ where ε is a small parameter with $\eta(x_1) = \eta(x_2) = 0$. We then have infinite paths passing through points (x_1, y_1) and (x_2, y_2).

$$\widehat{y} = y + \varepsilon\eta \tag{3.2}$$

where ε is a small parameter and η is a differentiable function satisfying

$$\eta(x_1) = \eta(x_1) = 0 \tag{3.3}$$

Over the path $\widehat{y}(x)$ we can write the functional

J.S. Rao, *Simulation Based Engineering in Solid Mechanics*,
DOI 10.1007/978-3-319-47614-8_3

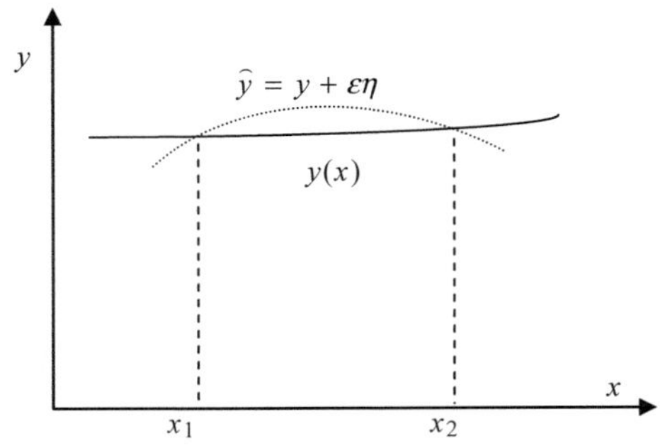

Fig. 3.1 Admissible and varied paths

$$\widehat{I} = \int_{x_1}^{x_2} F\left(x, \widehat{y}, \widehat{y}'\right)dx = \int_{x_1}^{x_2} F(x, y + \varepsilon\eta, y' + \varepsilon\eta')dx \tag{3.4}$$

We adopt the same strategy as in Calculus and expand the above functional as

$$\widehat{I} = \left(\widehat{I}\right)_{\varepsilon=0} + \left(\frac{\partial \widehat{I}}{\partial \varepsilon}\right)_{\varepsilon=0}\varepsilon + \left(\frac{\partial^2 \widehat{I}}{\partial \varepsilon^2}\right)_{\varepsilon=0}\frac{\varepsilon^2}{2!} + \cdots \tag{3.5}$$

$$\widehat{I} - \left(\widehat{I}\right)_{\varepsilon=0} = \widehat{I} - I = \left(\frac{\partial \widehat{I}}{\partial \varepsilon}\right)_{\varepsilon=0}\varepsilon + \left(\frac{\partial^2 \widehat{I}}{\partial \varepsilon^2}\right)_{\varepsilon=0}\frac{\varepsilon^2}{2!} + \cdots \tag{3.5a}$$

For $\widehat{I}$ to be an extremum, the first term on the right side must necessarily be zero,

$$\left(\frac{\partial \widehat{I}}{\partial \varepsilon}\right)_{\varepsilon=0} = 0 \tag{3.6}$$

$$\begin{aligned} &\therefore \left\{\int_{x_1}^{x_2} \frac{\partial}{\partial \varepsilon} F\left(x, \widehat{y}, \widehat{y}'\right)dx\right\}_{\varepsilon=0} = 0 \\ &\text{i.e.,} \left\{\int_{x_1}^{x_2} \left(\frac{\partial F}{\partial \widehat{y}}\frac{\partial \widehat{y}}{\partial \varepsilon} + \frac{\partial F}{\partial \widehat{y}'}\frac{\partial \widehat{y}'}{\partial \varepsilon}\right)dx\right\}_{\varepsilon=0} = 0 \end{aligned} \tag{3.7}$$

Or

$$\int_{x_1}^{x_2} \left(\frac{\partial F}{\partial y}\eta + \frac{\partial F}{\partial y'}\eta'\right)dx = 0 \tag{3.8}$$

Integrating second term in the above by parts and noting that η is zero at end points

$$\int_{x_1}^{x_2} \frac{\partial F}{\partial y'} \eta' dx = \cancel{\left. \frac{\partial F}{\partial y'} \eta \right|_{x_1}^{x_2}} - \int_{x_1}^{x_2} \frac{d}{dx}\left(\frac{\partial F}{\partial y'}\right) \eta' dx \tag{3.9}$$

Substituting Eq. (3.8) now becomes

$$\int_{x_1}^{x_2} \left[\frac{\partial F}{\partial y} - \frac{d}{dx}\left(\frac{\partial F}{\partial y'}\right)\right] \eta dx = 0 \tag{3.9a}$$

Since η is arbitrary, we have

$$\frac{d}{dx}\left(\frac{\partial F}{\partial y'}\right) - \frac{\partial F}{\partial y} = 0 \tag{3.10}$$

This is Euler-Lagrange Equation for a functional $F(x, y, y')$ with only first order derivative which gives the condition required for $y(x)$ to be an extremizing function.

Now let us consider the functional $F(x, y, y', y'')$ with second order derivative included that would represent a beam problem.

$$I = \int_{x_1}^{x_2} F(x, y, y', y'') dx \tag{3.11}$$

The independent variable is limited to the range x_1 to x_2 and we wish to find how the dependent variable varies in this range. We will simply follow the above steps

$$\left(\frac{d\hat{I}}{d\varepsilon}\right)_{\varepsilon=0} = \left[\frac{\partial}{\partial \varepsilon} \int_{x_1}^{x_2} F\left(x, \hat{y}, \hat{y}', \hat{y}''\right) dx\right]_{\varepsilon=0} = 0 \tag{3.12}$$

$$\left\{\int_{x_1}^{x_2} \left(\frac{\partial F}{\partial \hat{y}} \eta + \frac{\partial F}{\partial \hat{y}'} \eta' + \frac{\partial F}{\partial \hat{y}''} \eta''\right) dx\right\}_{\varepsilon=0} = 0 \tag{3.13}$$

$$\int_{x_1}^{x_2} \left(\frac{\partial F}{\partial y} \eta + \frac{\partial F}{\partial y'} \eta' + \frac{\partial F}{\partial y''} \eta''\right) dx = 0 \tag{3.14}$$

Integrating by parts, the middle term is

$$\int_{x_1}^{x_2} \frac{\partial F}{\partial y'} \eta' dx = \frac{\partial F}{\partial y'} \eta \bigg|_{x_1}^{x_2} - \int_{x_1}^{x_2} \frac{d}{dx}\left(\frac{\partial F}{\partial y'}\right) \eta dx \tag{a}$$

Similarly the last term upon integration by parts is

$$\int_{x_1}^{x_2} \frac{\partial F}{\partial y''} \eta'' dx = \frac{\partial F}{\partial y''} \eta' \bigg|_{x_1}^{x_2} - \int_{x_1}^{x_2} \frac{d}{dx}\left(\frac{\partial F}{\partial y''}\right) \eta' dx$$

$$= \frac{\partial F}{\partial y''} \eta' \bigg|_{x_1}^{x_2} - \frac{d}{dx}\left(\frac{\partial F}{\partial y''} \eta\right) \bigg|_{x_1}^{x_2} + \int_{x_1}^{x_2} \frac{d^2}{dx^2}\left(\frac{\partial F}{\partial y''}\right) \eta dx \tag{b}$$

Substituting in (3.14)

$$\int_{x_1}^{x_2} \left[\frac{d^2}{dx^2}\left(\frac{\partial F}{\partial y''}\right) \frac{d}{dx}\left(\frac{\partial F}{\partial y'}\right) + \frac{\partial F}{\partial y}\right] \eta dx$$
$$- \frac{\partial F}{\partial y'} \eta \bigg|_{x_1}^{x_2} + \frac{\partial F}{\partial y''} \eta' \bigg|_{x_1}^{x_2} - \frac{d}{dx}\left(\frac{\partial F}{\partial y'}\right) \eta \bigg|_{x_1}^{x_2} = 0 \tag{3.15}$$

In order to satisfy the above equation we have the Euler-Lagrange differential equation that should be satisfied, i.e., we seek a solution for this equation subjected to the condition that the definite integrals are also satisfied; the latter lead to boundary conditions. Summing up Euler-Lagrange gave us the most powerful tool for determining structural response to tractions or loads.

For a functional $F(x, y, y', y'')$ to be extremum between x_1 and x_2, we satisfy the differential equation

$$\frac{d^2}{dx^2}\left(\frac{\partial F}{\partial y''}\right) - \frac{d}{dx}\left(\frac{\partial F}{\partial y'}\right) + \frac{\partial F}{\partial y} = 0 \tag{3.16}$$

subject to Kinematic and Natural boundary conditions as given below.

Kinematic		Natural
y' specified (including set to zero)	or	$-\frac{\partial F}{\partial y''} = 0$
y specified (including set to zero)	or	$-\frac{d}{dx}\left(\frac{\partial F}{\partial y'}\right) - \frac{\partial F}{\partial y'} = 0$

We will illustrate how Euler-Lagrange equations are set up for simplest possible structures with a one dimensional idealization.

Note that these equations from Euler and Lagrange satisfy the universal principle that any physics process is conservative in nature. No energy can be destroyed nor generated. Energy can only transform from one form to another; in static structures the work done by traction loads during the deformation will be equal the strain energy stored in the system. The deformation field is however not known and without this we cannot determine the work or stored strain energy. Given a deformation field however, we can determine either the work done by the traction or the strain energy stored in the system. We want that deformation field that makes the work and stored energy to be equal. The genius of Euler and Lagrange lies in the fact of choosing that assumed deformation field which makes the difference between the work done and stored energy is minimum possible.

3.1 General Approach for Solving Structural Problems

In Chap. 2 we discussed the behavior of a solid under traction (or load) describing the deformations, strains and stresses suffered by the solid in this process. They form the starting point of any structural design. We restrict ourselves here to one dimensional structure, structures whose behavior is determined with only one independent variable x. The chapters that follow will discuss these structures in detail. As we mentioned there is no exact method of setting a problem and solving it. All physical problems begin with some assumptions to enable us express them mathematically. We begin with the following steps.

1. Assume a suitable displacement field u_x, u_y and u_z for the medium of the structure as a function of the distance x from the left end
2. Determine the strain field of the structure using

$$\begin{bmatrix} \varepsilon_{xx} & \varepsilon_{xy} & \varepsilon_{xz} \\ \varepsilon_{yx} & \varepsilon_{yy} & \varepsilon_{yz} \\ \varepsilon_{zx} & \varepsilon_{zy} & \varepsilon_{zz} \end{bmatrix} \quad \varepsilon_{ij} = \frac{1}{2}\left(\frac{\partial u_i}{\partial x_j} + \frac{\partial u_j}{\partial x_i}\right) \qquad (2.36)\text{ and }(2.37)$$

$$i = x, y, z; \quad j = x, y, z$$

3. Use Hooke's law to determine the stress field

$$\sigma_{ij} = \lambda \delta_{ij} \varepsilon_{ii} + 2G\varepsilon_{ij}$$

where

$$G = \frac{E}{2(1+\nu)}; \quad \lambda = \frac{E\nu}{(1+\nu)(1-2\nu)}$$
or
$$E = \frac{G(3\lambda + 2G)}{\lambda + G}; \quad \nu = \frac{\lambda}{2(\lambda + G)} \qquad (2.45)\text{ and }(2.46)$$

If the change in the cross-sectional area is ignored, we set $v \to 0$

$$\lambda + 2G = E \tag{2.47}$$

4. Determine the strain energy due to deformation in the structure by using

$$U = \iiint_v U_0 dv = \iiint_v \left[\int_0^{\varepsilon_{ij}} \sigma_{ij} d\varepsilon_{ij} \right] dv = \frac{1}{2} \iiint_v \sigma_{ij}\varepsilon_{ij} dv \tag{2.50}$$

5. Determine the work of external forces, W.
6. Set up the total potential energy given by $\pi = U - W$ and identify the functional to be used in Euler-Lagrange equation.
7. Derive the governing differential equation and identify the natural and kinematic boundary conditions; Solve the problem.

We will illustrate this procedure for a one-dimensional structure with three different load applications in the next three chapters.

Before we consider a structure let us consider a common problem that we find in our backyard at homes; a stretched string. A string can form a structure, in sense it can be made to carry a transverse load $Q(x)$ by stretching it first under an axial tension P. Without this axial tension in the string it cannot carry any load. Let $w(x)$ be the deflection at distance x of this string of length l, see Fig. 3.2.

Consider the element dx which undergoes a stretch with a final length $\sqrt{dw^2 + dx^2}$. The change in an elemental length dx is then

$$\Delta = \sqrt{dw^2 + dx^2} - dx = \frac{1}{2} w_{,x}^2 \, dx \tag{3.17}$$

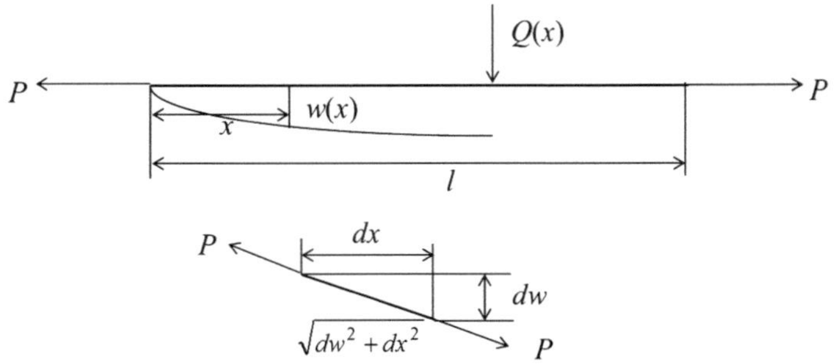

Fig. 3.2 A stretched string

Due to this stretch, the strain energy stored in the string is

$$U = \int_0^l \frac{1}{2} P w_{,x}^2 \, dx \tag{3.18}$$

The work done by external force $Q(x)$ is

$$W = \int_0^l Q(x) w dx \tag{3.19}$$

The Euler-Lagrange equation involving first derivatives alone is

$$\frac{d}{dx}\left(\frac{\partial F}{\partial y'}\right) - \frac{\partial F}{\partial y} = 0 \tag{3.10}$$

where the functional $F(x, y, y')$ is (take care of appropriate notations)

$$F(x, w, w') = \int_0^l \left(\frac{1}{2} P w_{,x}^2 - Q w\right) dx = \int_0^l F dx \tag{3.20}$$

$$F(x, w, w') = \frac{1}{2} P w_{,x}^2 - Q w \tag{3.21}$$

We have the following

$$\begin{aligned} \frac{\partial F}{\partial w'} &= P w_{,x} \\ \frac{d}{dx}\left(\frac{\partial F}{\partial w'}\right) &= P w_{,xx} \\ -\frac{\partial F}{\partial w} &= Q \end{aligned} \tag{3.22}$$

Substituting in Lagrange equation (3.10)

$$P w_{,xx} + Q = 0 \tag{3.23}$$

Rewriting the governing differential equation

$$\frac{d^2 w}{dx^2} = -\frac{Q}{P}$$

Integrating

$$\frac{dw}{dx} = -\frac{Q}{P}x + A$$
$$w = -\frac{Q}{P}\frac{x^2}{2} + Ax + B$$

The end (boundary) conditions are $w = 0$ at $x = 0$ and $x = l$. The first condition at $x = 0$ gives $B = 0$ and the second condition at $x = l$ gives

$$0 = -\frac{Q}{P}\frac{l^2}{2} + Al$$
$$\therefore A = \frac{1}{2}\frac{Ql}{P}$$

The deflection $w(x)$ is therefore

$$w = \frac{1}{2}\frac{Q}{P}x(l - x)$$

At midpoint $x = 1/2\ l$

$$w = \frac{Q}{P}\frac{l^2}{4}$$

If there is no tension i.e., $P = 0$ the deflection is infinity or rather indeterminate. Therefore strings by themselves without any tension cannot act as structural members.

Besides being useful for home applications, strings also form structural elements, e.g., in suspension bridges, see Fig. 3.3.

Fig. 3.3 A suspension bridge with strings (*Bars*) carrying the vehicular loads

3.2 Other Applications of Euler-Lagrange Equation Leading to Optimization

The extremization principle is fundamental to many aspects of Physics; one such problem is posed early during the scientific revolution. The Brachistochrone problem is specifying the curve connecting two points displaced from each other laterally, along which a body, acted upon only by gravity, would fall in the shortest time.

Let the two points be denoted by (1) and (2). Let the speed of the body be V and s is the total distance along the curve $y = f(x)$ as shown in Fig. 3.4. The time for descent from point (1), (x_1, y_1) to point (2) (x_2, y_2) is given by

$$t = \int_{(1)}^{(2)} \frac{dx}{V}$$

$$= \int_{(1)}^{(2)} \frac{\sqrt{dx^2 + dy^2}}{V}$$

$$= \int_{(1)}^{(2)} \frac{\sqrt{1 + \left(\frac{dy}{dx}\right)}}{V} dx \tag{3.24}$$

If V_1 is the speed at point (1), (x_1, y_1), then at any position (x, y) during the travel, we have

$$\frac{1}{2} m V_1^2 + m g y_1 = \frac{1}{2} m V^2 + m g y$$

Therefore, the speed at any position (x, y) is

$$V = \sqrt{V_1^2 - 2g(y - y_1)}$$

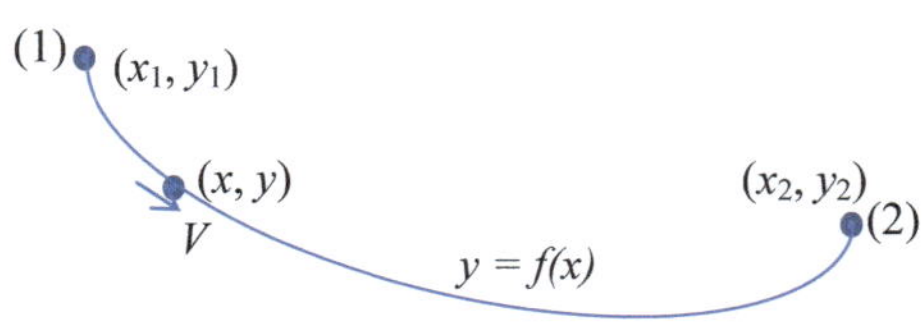

Fig. 3.4 Brachistochrone problem

Substituting the above in (3.24), we get

$$t = \int_{x_1}^{x_2} \frac{\sqrt{1 + \left(\frac{dy}{dx}\right)^2}}{\sqrt{V_1^2 - 2g(y - y_1)}} dx \tag{3.25}$$

For the special case $y_1 = 0$ and $V_1 = 0$, we get

$$t = \int_{x_1}^{x_2} \sqrt{\frac{1 + \left(\frac{dy}{dx}\right)^2}{2gy}} dx \tag{3.25a}$$

In optimization terminology, the above represents the Objective Function that is required to be minimized subject to constraints on starting and finishing points and the initial velocity. We can use Euler-Lagrange equation for this purpose.

$$\frac{\partial F}{\partial y} - \frac{d}{dx}\left(\frac{\partial F}{\partial y'}\right) = 0 \tag{3.10}$$

that makes t an extremum. Therefore the function F for this problem is

$$F(x, y, y') = \left(\frac{1 + \left(\frac{dy}{dx}\right)^2}{2gy}\right)^{\frac{1}{2}} \tag{3.26}$$

Using the above in Euler-Lagrange equation (3.10), we get

$$2y'' + y'^2 + 1 = 0 \tag{3.27}$$

The above is an ordinary nonlinear differential equation, which is not normally amenable for a closed form solution. In this case, however, we get the solution in parametric form as

$$\begin{aligned} x &= \frac{1}{2} C_1(\theta - \sin\theta) \\ y &= \frac{1}{2} C_1(1 - \cos\theta) \end{aligned} \tag{3.28}$$

These two equations represent the desired curve for Brachistochrone problem, for the least travel of time between two given points. The curve is a cycloid as shown in Fig. 3.5. For the solution to pass through ($x_2 = 10$, $y_2 = 10$)

Fig. 3.5 Cycloid

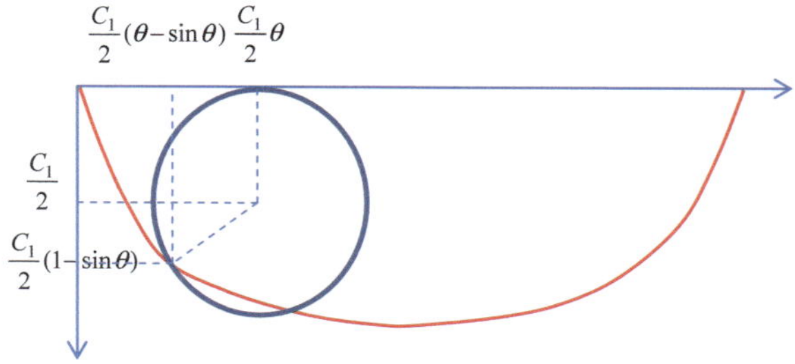

$$\begin{aligned} 10 &= \frac{1}{2}C_1(\theta - \sin\theta) \\ 10 &= \frac{1}{2}C_1(1 - \cos\theta) \end{aligned} \tag{3.29}$$

Subtracting the second one from the first of equations in (3.29)

$$(\theta - 1 - \sin\theta + \cos\theta) = 0$$

Solving

$$\theta = 2.412$$

Then,

$$\begin{aligned} \frac{1}{2}C_1 &= \frac{10}{(1 - \cos 2.412)} \\ &= 5.7292 \end{aligned}$$

This gives us the equation for cycloid of the Brachistochrone problem in the required range $0 \le x \le 10$ $0 \le y \le 10$.

$$\begin{aligned} x &= 5.7292(\theta - \sin\theta) \\ y &= 5.7292(1 - \cos\theta) \end{aligned} \tag{3.30}$$

From Eq. (3.30) the cycloid is given in Table 3.1.

The data in Table 3.1 obtained from Euler-Lagrange equation is plotted in Fig. 3.6 along with a straight line path from (0, 0) to (10, −10).

We note that the body initially begins the descent almost vertically that allows a maximum acceleration and thus complete the descent in minimum time.

Worked Example 3.1

We have shown earlier that a function or functional can have an extremum and Euler-Lagrangian equation gives the condition of these extrema. Let the functional be $F(x\ y, y') = x^2 + 2xy'^2 + xy$. Using Euler-Lagrangian Eq. (3.10)

Table 3.1 Cycloid in Eq. (3.30)

0	0
0.000954389	−0.028622136
0.00762367	−0.114202562
0.025665632	−0.255886187
0.060624433	−0.452257353
0.117875204	−0.701353986
0.202570341	−1.000687195
0.319588026	−1.347266141
0.473483484	−1.737627921
0.668443469	−2.16787217
0.908244434	−2.633700029
1.196214793	−3.130457101
1.535201669	−3.653179953
1.927542444	−4.196645711
2.375041407	−4.755424245
2.878951723	−5.323932424
3.439962913	−5.896489903
4.058193968	−6.467375877
4.733192153	−7.030886241
5.463937538	−7.581390586
6.248853182	−8.113388456
7.085820897	−8.621564302
7.972202403	−9.100840601
8.904865698	−9.546428581
9.880216346	−9.953876075
9.999940269	−10.00000915

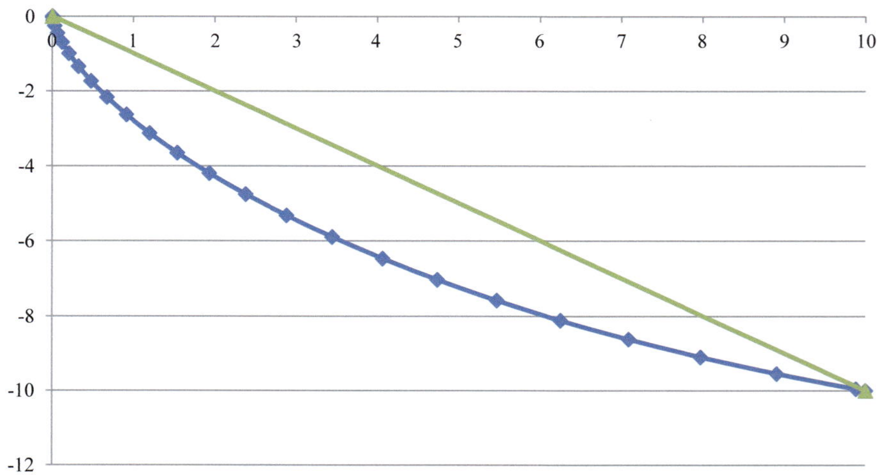

Fig. 3.6 Cycloid compared to straight line

$$\frac{\partial F}{\partial y} - \frac{d}{dx}\left(\frac{\partial F}{\partial y'}\right) = 0$$

$$\therefore x - \frac{d}{dx}(4xy') = 0$$

$$\text{i.e., } x - 4xy'' - 4y' = 0$$

The required Euler-Lagrangian equation is $4xy'' + 4y' - x = 0$.

Worked Example 3.2

Let the functional be $F(x, y, y') = x^2 + 2xy' + 1$. Then

$$\frac{\partial F}{\partial y} - \frac{d}{dx}\left(\frac{\partial F}{\partial y'}\right) = -2$$

Therefore, Euler-Lagrangian equation cannot be satisfied. Or, there is no $y(x)$ that extremizes the integral $I = \int_{x_1}^{x_2} (x^2 + 2xy' + 1)dx$ in (3.1).

3.3 Derivation of Euler-Lagrange Equation Through Delta Operator

Referring to Fig. 3.1, let us define an operator δ, called the delta operator and define $\delta[y(x)]$, delta of function $y(x)$ as

$$\delta[y(x)] = \hat{y}(x) - y(x) \tag{3.31}$$

δy is a variation of y. Note that we do not associate a δx with each δy. In a differentiation process a dy is associated with a given dx as shown in Fig. 3.7. δy is merely the vertical distance between points on different admissible paths at the same value of x whereas dy is the vertical distance between points on the same curve at positions dx apart.

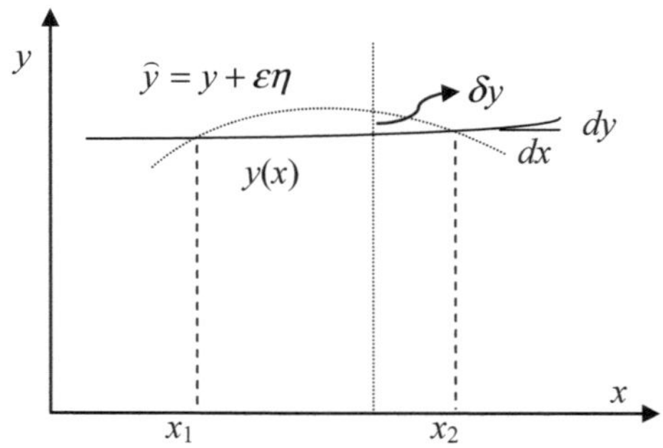

Fig. 3.7 Variation and differentiation illustrated

Following the above definition of variation of $y(x)$ we can write the variation of $\frac{dy}{dx}$

$$\delta\left[\frac{dy}{dx}\right] = \frac{d\hat{y}}{dx} - \frac{dy}{dx} = \frac{d}{dx}(\hat{y} - y) = \frac{d(\delta y)}{dx} \tag{3.32}$$

Therefore the δ operator is commutative with the differential operator d. Also $\int \hat{y}(x)dx$ is a varied function for $\int y(x)dx$ therefore delta operator is commutative with the integral operator. Equation (3.2) can be rewritten as

$$\delta y = \widehat{y} - y = \varepsilon\eta \tag{3.32a}$$

and

$$\delta y' = \widehat{y}' - y' = \varepsilon\eta' \tag{3.32b}$$

The functional F in Eq. (3.4) can be expressed using delta operator

$$F(x, y + \delta y, y' + \delta y')$$

Expanding the functional F as a Taylor series about y and y'

$$\begin{aligned} F(x, y + \delta y, y' + \delta y') &= F(x, y, y') + \left(\frac{\partial F}{\partial y}\delta y + \frac{\partial F}{\partial y'}\delta y'\right) + O(\delta^2) \Rightarrow \\ F(x, y + \delta y, y' + \delta y') - F(x, y, y') &= \left(\frac{\partial F}{\partial y}\delta y + \frac{\partial F}{\partial y'}\delta y'\right) + O(\delta^2) \end{aligned} \tag{3.33}$$

The left side of the above equation represents the total variation of the functional $\delta^{(T)}(F)$. The first term on the right side of the equation is called the first variation $\delta^{(1)}(F)$. We here afterwards neglect all the higher order terms, therefore

$$\begin{aligned} \delta^{(T)}F &= F(x, y + \delta y, y' + \delta y') - F(x, y, y') \\ \delta^{(1)}F &= \left(\frac{\partial F}{\partial y}\delta y + \frac{\partial F}{\partial y'}\delta y'\right) \end{aligned} \tag{3.34}$$

Integrating (3.33) between x_1 and x_2

$$\int_{x_1}^{x_2} F(x, y + \delta y, y' + \delta y')dx - \int_{x_1}^{x_2} F(x, y, y')dx = \int_{x_1}^{x_2} \left(\frac{\partial F}{\partial y}\delta y + \frac{\partial F}{\partial y'}\delta y'\right)dx \tag{3.35}$$

Or

$$\delta^{(T)}I = \hat{I} - I = \delta^{(1)}I \tag{3.36}$$

We can integrate the second term in the right-hand expression in (3.35) by parts and get

$$\int_{x_1}^{x_2} \left(\frac{\partial F}{\partial y'}\delta y'\right)dx = \frac{\partial F}{\partial y'}\delta y\Bigg|_{x_1}^{x_2} - \int_{x_1}^{x_2} \frac{d}{dx}\left(\frac{\partial F}{\partial y'}\right)\delta y dx \tag{3.37}$$

Substituting the above result in (3.35) we get the first variation as

$$\delta^{(1)}I = \int_{x_1}^{x_2} \left(\frac{\partial F}{\partial y} - \frac{d}{dx}\frac{\partial F}{\partial y'}\right)\delta y dx + \frac{\partial F}{\partial y'}\delta y\Bigg|_{x_1}^{x_2} \tag{3.38}$$

Since $\delta y = 0$ at x_1 and x_2, the definite integral vanishes leaving the first variation to be

$$\delta^{(1)}I = \int_{x_1}^{x_2} \left(\frac{\partial F}{\partial y} - \frac{d}{dx}\frac{\partial F}{\partial y'}\right)\delta y dx \tag{3.38a}$$

For I to be a maximum or a minimum its sign should be same for all possible variations δy over the interval x_1 to x_2. Thus δy at any position x could be $\pm$d where d is a small number. Therefore the bracketed expression in the integrand on the right side of the Eq. (3.35) must be zero. This leads to Eq. (3.38a) giving the familiar Euler-Lagrange equation derived earlier.

Worked Example 3.3

Derive using Delta operator, Euler-Lagrange equations for functionals with higher order derivatives given by

$$I = \int_{x_1}^{x_2} F(x, y, y', y'', y''')dx \tag{a}$$

As before we assume one-parameter family of admissible functions for extremizing I, then Eq. (3.12) gives

$$\left(\frac{d\hat{I}}{d\varepsilon}\right)_{\varepsilon=0} = \left[\frac{d}{d\varepsilon}\int_{x_1}^{x_2} F(x, \hat{y}, \hat{y}', \hat{y}'', \hat{y}''')dx\right]_{\varepsilon=0} = 0 \tag{b}$$

Following the steps given at the beginning of this chapter, the above leads to (3.13)

$$\left\{\int_{x_1}^{x_2}\left(\frac{\partial F}{\partial \hat{y}}\eta+\frac{\partial F}{\partial \hat{y}'}\eta'+\frac{\partial F}{\partial \hat{y}''}\eta''+\frac{\partial F}{\partial \hat{y}'''}\eta''\right)dx\right\}_{\varepsilon=0}=0$$
$$\Rightarrow\int_{x_1}^{x_2}\left(\frac{\partial F}{\partial y}\eta+\frac{\partial F}{\partial y'}\eta'+\frac{\partial F}{\partial y''}\eta''+\frac{\partial F}{\partial y''}\eta''\right)dx=0 \tag{c}$$

We can evaluate the above integrals in Eq. (c) by parts as shown below:

$$\int_{x_1}^{x_2}\left(\frac{\partial F}{\partial y'}\eta'\right)dx=\frac{\partial F}{\partial y'}\eta\Big|_{x_1}^{x_2}-\int_{x_1}^{x_2}\frac{d}{dx}\left(\frac{\partial F}{\partial y'}\right)\eta dx$$
$$\int_{x_1}^{x_2}\left(\frac{\partial F}{\partial y''}\eta''\right)dx=\frac{\partial F}{\partial y''}\eta'\Big|_{x_1}^{x_2}-\int_{x_1}^{x_2}\frac{d}{dx}\left(\frac{\partial F}{\partial y''}\right)\eta' dx$$
$$=\frac{\partial F}{\partial y''}\eta'\Big|_{x_1}^{x_2}-\frac{d}{dx}\left(\frac{\partial F}{\partial y''}\right)\eta\Big|_{x_1}^{x_2}+\int_{x_1}^{x_2}\frac{d^2}{dx^2}\left(\frac{\partial F}{\partial y''}\right)\eta dx$$
$$\int_{x_1}^{x_2}\left(\frac{\partial F}{\partial y'''}\eta'''\right)dx=\frac{\partial F}{\partial y'''}\eta''\Big|_{x_1}^{x_2}-\int_{x_1}^{x_2}\frac{d}{dx}\left(\frac{\partial F}{\partial y'''}\right)\eta'' dx$$
$$=\frac{\partial F}{\partial y'''}\eta''\Big|_{x_1}^{x_2}-\frac{d}{dx}\left(\frac{\partial F}{\partial y'''}\right)\eta'\Big|_{x_1}^{x_2}+\int_{x_1}^{x_2}\frac{d^2}{dx^2}\left(\frac{\partial F}{\partial y'''}\right)\eta' dx$$
$$=\frac{\partial F}{\partial y'''}\eta''\Big|_{x_1}^{x_2}-\frac{d}{dx}\left(\frac{\partial F}{\partial y'''}\right)\eta'\Big|_{x_1}^{x_2}+\frac{d^2}{dx^2}\left(\frac{\partial F}{\partial y'''}\right)\eta\Big|_{x_1}^{x_2}-\int_{x_1}^{x_2}\frac{d^3}{dx^3}\left(\frac{\partial F}{\partial y'''}\right)\eta dx \tag{d}$$

Putting them together, Eq. (c) becomes

$$-\int_{x_1}^{x_2}\left[\frac{d^3}{dx^3}\left(\frac{\partial F}{\partial y'''}\right)-\frac{d^2}{dx^2}\left(\frac{\partial F}{\partial y''}\right)+\frac{d}{dx}\left(\frac{\partial F}{\partial y'}\right)-\frac{\partial F}{\partial y}\right]\eta dx$$
$$+\frac{\partial F}{\partial y'''}\eta''\Big|_{x_1}^{x_2}-\left\{\frac{d}{dx}\left(\frac{\partial F}{\partial y'''}\right)-\frac{\partial F}{\partial y''}\right\}\eta'\Big|_{x_1}^{x_2}$$
$$+\left\{\frac{d^2}{dx^2}\left(\frac{\partial F}{\partial y'''}\right)-\frac{d}{dx}\left(\frac{\partial F}{\partial y''}\right)+\frac{\partial F}{\partial y'}\right\}\eta\Big|_{x_1}^{x_2}=0 \tag{e}$$

Since η, η', η'' are zero at both $x = x_1$ and x_2, all the definite integrals vanish, leaving

$$-\int_{x_1}^{x_2}\left[\frac{d^3}{dx^3}\left(\frac{\partial F}{\partial y'''}\right)-\frac{d^2}{dx^2}\left(\frac{\partial F}{\partial y''}\right)+\frac{d}{dx}\left(\frac{\partial F}{\partial y'}\right)-\frac{\partial F}{\partial y}\right]\eta dx = 0 \tag{f}$$

The Euler-Lagrange equation for (a) then is given by

$$\frac{d^3}{dx^3}\left(\frac{\partial F}{\partial y'''}\right)-\frac{d^2}{dx^2}\left(\frac{\partial F}{\partial y''}\right)+\frac{d}{dx}\left(\frac{\partial F}{\partial y'}\right)-\frac{\partial F}{\partial y} = 0 \tag{g}$$

η, η', η'' can be either zero at both $x = x_1$ and x_2 or they can be specified and these conditions are called *kinematic boundary conditions*. To have an extremal condition of the integral in (a), we can satisfy the kinematic boundary conditions in (e) and the Euler-Lagrange equation given in (g); alternatively we can satisfy the second set of conditions that come out naturally in (e), viz.,

$$\frac{\partial F}{\partial y'''} = 0;\quad \left\{\frac{d}{dx}\left(\frac{\partial F}{\partial y'''}\right)-\frac{\partial F}{\partial y''}\right\} = 0;$$
$$\left\{\frac{d^2}{dx^2}\left(\frac{\partial F}{\partial y'''}\right)-\frac{d}{dx}\left(\frac{\partial F}{\partial y''}\right)+\frac{\partial F}{\partial y'}\right\} = 0 \quad \text{at } x = x_1 \text{ and } x = x_2 \tag{h}$$

Since these conditions come out naturally in the derivation of Euler-Lagrange equation, they are called *Natural Boundary Conditions.* Summarizing.

We can satisfy at the boundaries any of the three boundary conditions given in Table Worked Example 3.3 and solve the Euler-Lagrange equation. The greatest advantage of the Euler-Lagrange and Calculus of Variations approach lies in the fact that all boundary conditions are a part of the derivation. The natural boundary conditions reflect the forces and kinematic boundary conditions are related to the path, in structural problems, the deflections and slopes of the deflection curve.

Note that by disregarding the terms involving y''' we get the results for a functional whose highest-order derivative is second-order. Similarly by disregarding both terms with y''' and y'', we get back to the case where y' is the highest-order term in the functional.

Table Worked Example 3.3 Kinematic and natural boundary conditions

Kinematic	Natural
y'' specified	$\frac{\partial F}{\partial y'''} = 0$
y' specified	$\left\{\frac{d}{dx}\left(\frac{\partial F}{\partial y'''}\right)-\frac{\partial F}{\partial y''}\right\} = 0$
y specified	$\left\{\frac{d^2}{dx^2}\left(\frac{\partial F}{\partial y'''}\right)-\frac{d}{dx}\left(\frac{\partial F}{\partial y''}\right)+\frac{\partial F}{\partial y'}\right\} = 0$

3. Exercises

3.1 Fermat's principle states that the path of light is such as to minimize the time passage from one point to another. Determine the path of light y in a plane medium between two points wherein the speed of light is given as Cy.

3.2 Given the functional $I = \int_{x_1}^{x_2} F(x, y'''')dx$ determine its Euler-Lagrange equation and natural boundary conditions.

3.3 What are admissible and varied paths? Using this approach derive Euler-Lagrange equations.

3.4 Compare extremal conditions of a function and a functional and show their similarities.

3.5 What is a delta operator? Using a delta operation determine the Euler-Lagrange equations.

3.6 What are natural and kinematic boundary conditions? How do you obtain them?

3.7 Why are boundary conditions necessary to obtain the extremal conditions of a given functional?

3.8 Can you solve a structural problem just by using Euler-Lagrange equation alone? Explain?

3.9 Show how you choose the boundary conditions from amongst the natural and kinematic conditions?

3.10 Outline the general approach of solving a structural problem from Euler-Lagrange equations. Why is it that you have to start with an assumed solution?

Chapter 4
Axially Loaded 1-D Structures

There are no simple structures; all structures are complex and three-dimensional in nature with varying geometry as we have seen in Chap. 1. We have to find ways and means of understanding these structures and find a solution for their deformations, strains and stresses as defined in Chap. 2 following methods such as the one described in Chap. 3. We have to adopt some method of numerical computation to achieve this for a real life structure that may be stationary as in a building, temple, bridge, foundation, casing of a machine etc. or a rotating structure such as a shaft in a turbine or generator. Over the last century, processes were developed depending on the available computational means at a given time. Initially they are graphical methods, tabular methods using log tables for numerical calculations, followed by slide rules or mechanical calculators then with calculators; today it is mostly achieved by computers initially with programs developed in FORTRAN or ALGOL followed by dedicated commercial software which can churn around results faster for very complex problems. We will work as we go along in this book to introduce the methods that will lead to modern computer methods.

First we define an ideal structure and the assumptions made in deriving such a structure. The simplest of the structures is a one dimensional structure. Of course there are no one-dimensional structures, we need to use some solids to build a structure, e.g., a three-dimensional brick is used in the construction of a wall or beam or column. When we want to model such a structure we can identify individual members, e.g., we have several columns in a multistory building or several columns in the foundation of a high powered turbo-generator set. These structural members are usually long when compared to the cross-sectional dimensions.

Consider a bar of uniform cross-section of area A with uniform axial load P N/m applied along the x axis of length l as shown in Fig. 4.1.

J.S. Rao, *Simulation Based Engineering in Solid Mechanics*,
DOI 10.1007/978-3-319-47614-8_4

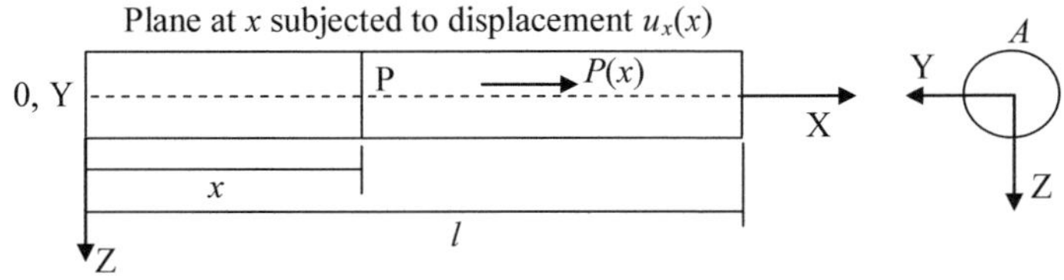

Fig. 4.1 Axially loaded bar

We follow the steps outlined in the previous Chap. 3.

1. Displacement field of a point P at distance x

$$\begin{aligned} u_x &= u(x) \\ u_y &= 0 \\ u_z &= 0 \end{aligned} \tag{4.1}$$

Strictly speaking the above field is incorrect, since from theory of elasticity we know that displacement in x direction will be accompanied by Poisson displacements in y and z directions. For slender bars this assumption is valid that allows us to keep the theory simple with only one independent variable x.

2. Obtain the Strain field

$$\varepsilon_{ij} = \frac{1}{2}\left(\frac{\partial u_i}{\partial x_j} + \frac{\partial u_j}{\partial x_i}\right) \Rightarrow \varepsilon_{xx} = u_{,x} \text{ rest } \varepsilon = 0 \tag{4.2}$$

3. Obtain the Stress field by ignoring the Poisson's ratio

$$\sigma_{ij} = \lambda\delta_{ij}\varepsilon_{ii} + 2G\varepsilon_{ij} \Rightarrow \sigma_{xx} = Eu_{,x} \text{ rest } \sigma = 0 \tag{4.3}$$

4. Strain Energy of the Bar

$$U = \frac{1}{2}\iiint\limits_v \sigma_{ij}\varepsilon_{ij} dv = \frac{1}{2}\int\limits_0^l EAu_{,x}^2\, dx \tag{4.4}$$

5. The work done by external force for the one dimensional bar

$$W = \int\limits_0^l Pu dx \tag{4.5}$$

6. Total Potential Energy and Functional

$$\pi = U - W = \int_0^l \left(\frac{1}{2}EAu_{,x}^2 - Pu\right)dx = \int_0^l F(u, u', x)dx$$
$$\Rightarrow F(u, u', x) = \frac{1}{2}EAu_{,x}^2 - Pu \tag{4.6}$$

7. Euler-Lagrange Equation and Governing Differential Equation and Boundary Conditions

$$-\frac{d}{dx}\left(\frac{\partial F}{\partial y'}\right) + \frac{\partial F}{\partial y} = 0$$
$$\frac{\partial F}{\partial u'} = EAu_{,x}$$
$$-\frac{d}{dx}\left(\frac{\partial F}{\partial u'}\right) = -EAu_{,xx}$$
$$\frac{\partial F}{\partial u} = -P$$

$$-EAu_{,xx} = P \quad \text{i.e.,} \quad -EA\frac{d^2u}{dx^2} = P \tag{4.7}$$

$$u = 0 \quad \text{or} \quad u_{,x} = 0 \quad \text{at}\, x = 0 \,\text{and}\, x = l \tag{4.8}$$

4.1 Simply Supported Bar

Now consider a uniform bar of cross-section area A with both ends simply supported or hinged. The governing differential equation is

$$\frac{d^2u}{dx^2} = -\frac{P}{EA} \tag{4.7a}$$

Integrating

$$\frac{du}{dx} = -\frac{P}{EA}x + M$$
$$u = -\frac{1}{2}\frac{P}{EA}x^2 + Mx + B \tag{4.9}$$

M and B are constants of integration and obtained from the boundary conditions.

$$u = 0 \quad \text{at} \quad x = 0\,\text{gives}\,B = 0 \quad \therefore u = -\frac{1}{2}\frac{P}{EA}x^2 + Mx$$

$$u = 0 \quad \text{at} \quad x = l \quad \text{gives}\,M = \frac{1}{2}\frac{P}{EA}l$$

Therefore, the longitudinal or axial deflection of the bar is

$$u = \frac{1}{2}\frac{P}{EA}x(l - x) \tag{4.10}$$

At either end the displacement is zero. The displacement at midpoint $x = \frac{1}{2}l$ is

$$u_{x=\frac{1}{2}l} = \frac{1}{8}\frac{Pl^2}{EA} \tag{4.11}$$

Let us normalize the x axis values by defining $X = \frac{x}{l}$; substituting $x = lX$ in (4.10)

$$u = \frac{1}{2}\frac{Pl^2}{EA}X(1 - X)$$

Table 4.1 gives the displacement at $X = 0, \frac{1}{4}, \frac{1}{2}, \frac{3}{4}$ and 1. We notice that the displacement increases gradually as a parabola from left end to a maximum at the middle and then decreases to zero again at the right end.

The strain at any point x is direct strain along the axis given by

$$\begin{aligned}
\varepsilon_{xx} &= \frac{du}{dx} = \frac{1}{2}\frac{P}{EA}\frac{d}{dx}\left(xl - x^2\right) \\
&= \frac{1}{2}\frac{P}{EA}(l - 2x) \\
&= \frac{1}{2}\frac{Pl}{EA}(1 - 2X)
\end{aligned} \tag{4.12}$$

The rest of the five strains, two direct and three shear are zero; this is not strictly correct because we made them zero by our assumptions. If we didn't make this approximation the problem will be more complicated. Engineers do not worry

Table 4.1 Axial deflection along the bar

X	$X(1 - X)$	$u/\frac{Pl^2}{EA}$
0	0	0
1/4	3/16	3/32
1/2	1/4	1/8
3/4	3/16	3/32
1	0	0

about non-exactness, because these non-zero strains are negligible when compared to the direct axial strain. The effort of keeping the exactness is not worth for the additional accuracy we may get. The simple assumption of (4.1) allows us to get a reasonably good solution. Remember that all elasticity problems start from an assumed shape to derive the governing equations.

We see that the strain at the mid-point is zero and is a maximum at either end; at the left end $\varepsilon_{xxx=0} = \frac{1}{2}\frac{Pl}{EA}$ the strain is positive and decreases from this maximum value to a zero value at the center of the bar, as the deflection is increasing from a zero value to a maximum value. From the center of the bar, the deflection begins to decrease and reach a zero value at the right support and the strain now negative in the second half of the bar correspondingly increases from zero value to a maximum value. At the end of right side $\varepsilon_{xxx=l} = -\frac{1}{2}\frac{Pl}{EA}$.

Likewise the stresses are (Strain multiplied by Young's modulus E)

$$\sigma_{xxx=0} = \frac{Pl}{2A}; \quad \sigma_{xxx=\frac{1}{2}l} = 0; \quad \sigma_{xxx=l} = -\frac{Pl}{2A} \tag{4.13}$$

Like strain, the stresses are tensile in the first half and become compressive in the second half.

4.2 Simply Supported—Free Bar

In this case, the right end is free and expands under the uniformly distributed load from left end. The slope at the right free end is zero. The left boundary condition is same as in the Simply Supported Bar, hence $B = 0$. At the right end $x = l$, the slope is

$$\left.\frac{du}{dx}\right\}_{x=l} = -\frac{P}{EA}l + M = 0 \Rightarrow M = \frac{Pl}{EA} \tag{4.14}$$

Therefore

$$\begin{aligned} u &= -\frac{1}{2}\frac{P}{EA}x^2 + \frac{Pl}{EA}x \\ u' &= -\frac{P}{EA}x + \frac{Pl}{EA} \end{aligned} \tag{4.15}$$

The axial or longitudinal deflection increases in a parabolic manner from zero to a maximum value of $\frac{Pl^2}{2EA}$.

At the left support the strain is a maximum with a value $\frac{Pl}{EA}$ which decreases in a linear manner to zero at the free end. The stress likewise is also a maximum at the left end $\frac{Pl}{A}$ and reaches zero at the free right end.

4.3 Finite Element Method

The structure in Fig. 4.1 is not only idealized as a one-dimensional structure in place of an actual structure like a pillar which is a three dimensional structure. Rarely such uniform structures occur in practice. A structure idealized by violating Poisson's law may be in practice a non-uniform structure and is considered in the class of one-dimensional structures.

Figure 4.2a shows a tapered bar and its area of cross-section can be expressed as a function of x. But in Fig. 4.2b we have a complex structure whose cross-sectional area is not amenable for expressing as a function of x. There can be stepped bars with area of cross-section changing at specified positions along the x axis. In the absence of present day computational facilities, these structures were dealt by equilibrium methods to determine average stresses at any cross-section by balancing the internal force given by average stress multiplied by area A with the external load.

Today we adopt a process called Finite Element Method, which is not entirely new but not implemented until recently with the availability of high performance computing. In this approach a structure is divided into several elements of finite dimensions however small they may be. They are not infinitesimally small elements whose dimensions tend to zero in deriving the properties at a point by employing calculus; instead each element is treated as a structure. They are *finite elements* or finite substructures that can be assembled through the points of interconnectivity called *nodes*. This assembly will provide the original structure for which analysis is required.

In the finite element method each element is treated as a substructure subjected to the same physics of the original structure, in these axially loaded bars. Let us illustrate first general notation by considering a plate simply supported and acted upon by two loads 50 N each as shown in Fig. 4.3a. This structure is divided into several substructures or finite elements as shown in Fig. 4.3b. Each element is given an ID as shown in Fig. 4.3b. These finite elements are interconnected and the points (nodes) of interconnectivity are also numbered as shown. Each of these points of interconnectivity, nodes, are also given an ID, i.e., given a nodal number.

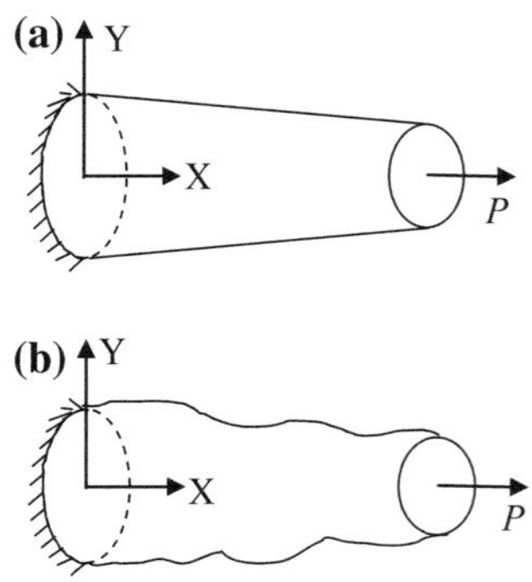

Fig. 4.2 Bars of variable cross-section

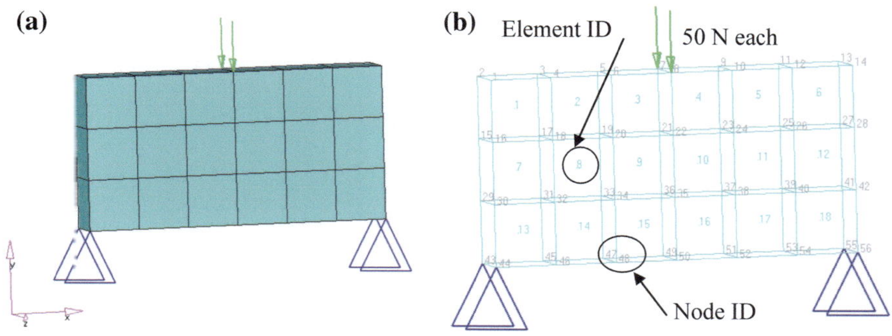

Fig. 4.3 Finite elements, nodes of a structure

Each element is treated as a structure or a domain on its own. Within each of these small domains, we assume a simple general solution to the governing equations which is a function of the unknown solutions at the nodes. The present problem of an axially loaded structure under consideration is shown in Fig. 4.4 with its solution as obtained for the continuously distributed system with infinite number of elements and nodes. Each node in general has one or more degrees of freedom to move in translation or rotation, therefore the domain has infinite degrees of freedom.

The nodal values in Fig. 4.4 are continuous with infinite degrees of freedom.

The fundamental concept of FEM is that the continuous function of a given domain Ω having infinite degrees of freedom is replaced by a discrete model, approximated by a set of piecewise continuous function having a finite degree of freedom as shown in Fig. 4.5a. The nodal values of the original continuous domain are compared with the finite element discrete nodal values in Fig. 4.5b. If the domain is divided into a fairly large number of elements we expect the nodal values obtained from such discretized domain are close to the exact solution. Through this way we can approach the type of structures which are not uniformly continuous as in a bar, rod, beam, plate, shell or solids. Here we restrict ourselves to one dimensional structure.

For an axially loaded one dimensional structure, let us first outline the finite element method. The total potential energy is given by $\pi = U - W$. We express

Fig. 4.4 Domain with infinite degrees of freedom

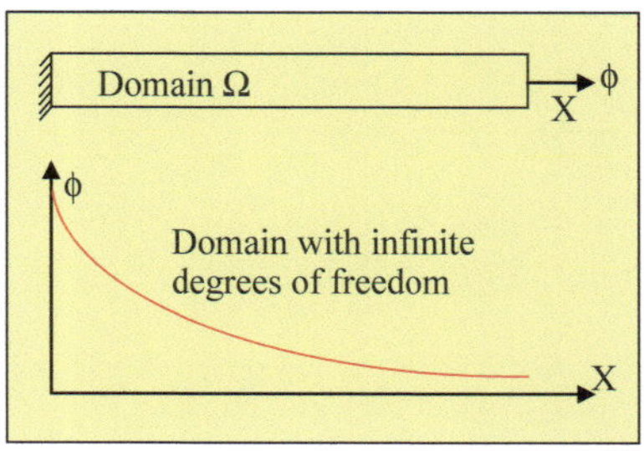

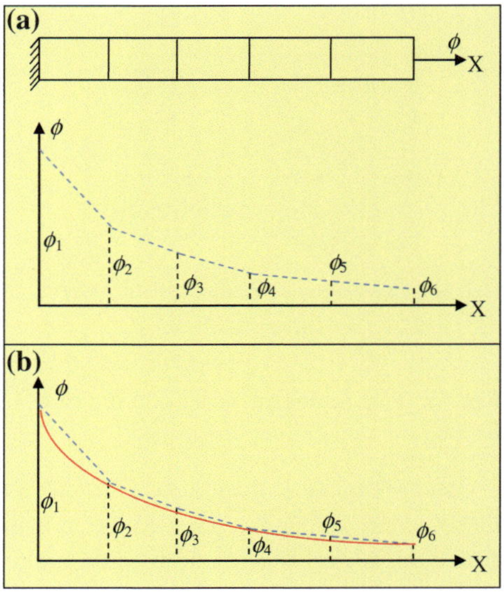

Fig. 4.5 Continuous domain divided into several sub-domains—nodal values

U to be a function of generalized coordinates given by the nodal degrees of freedom, i.e., U is a function of q. q in this case can be simply nodal axial displacements. We follow the energy principle and use the varied path approach, i.e., we will determine the variation of the total potential energy to be zero, expressed as

$$\delta \int_{0}^{L} \pi dx = \delta \int_{0}^{L} [U(q) - W]dx = 0 \tag{4.16}$$

The symbol δ denotes variation taken over the varied path q; i.e., variation of the integral taken over the domain for the total potential energy is set to zero to satisfy the condition that π is stationary. Taking the variation of different terms in the above, we get

$$\begin{aligned} \delta U &= \frac{\partial U}{\partial q} \delta q \\ \delta W &= P \delta q \end{aligned} \tag{4.17}$$

Substituting in (4.16)

$$\begin{aligned}\delta\int_0^L [U(q)-W]dx &= \int_0^L (\delta U - P\delta q)dx \\ &= \int_0^L \left(\frac{\partial U}{\partial q}\delta q - P\delta q\right)dx \\ &= \int_0^L \left(\frac{\partial U}{\partial q} - P\right)\delta q dx\end{aligned} \tag{4.18}$$

$$\therefore \frac{\partial U}{\partial q} = P \tag{4.19}$$

We follow the steps given below:

Step 1:

Choose the element, see Fig. 4.6. The element in Fig. 4.6 is element 1, followed by other elements that make up the structure. This element is defined in its own local coordinates denoted by $\widehat{X}\widehat{Y}\widehat{Z}$ and subjected to nodal forces $\widehat{f}_{1x}$ and $\widehat{f}_{2x}$ at the nodes 1 and 2 respectively. The displacements of the two nodes are assumed to be known during the formulation.

Step 2: Define Shape Function Matrix

The displacement at any location within the element is obtained by assuming a shape function or interpolation function. Here there are only two nodes and their displacement values are $\widehat{u}_1$ and $\widehat{u}_2$. To satisfy these conditions we assume a linear shape function for the displacement $\widehat{u}\left(\widehat{x}\right)$.

$$\widehat{u}\left(\widehat{x}\right) = a_1 + a_2\widehat{x} \tag{4.20}$$

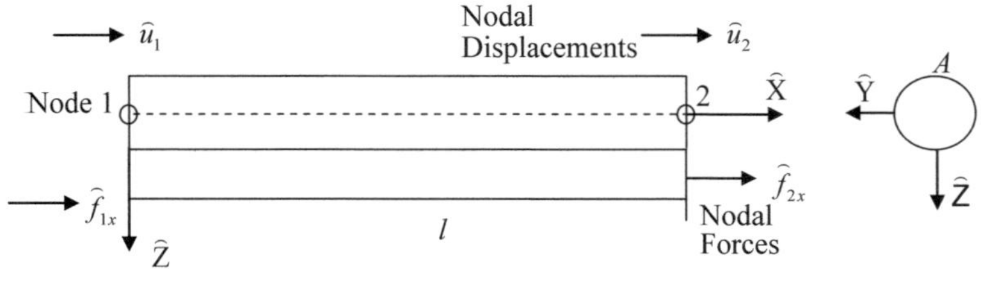

Fig. 4.6 Element 1 in local coordinates $\widehat{X}\widehat{Y}\widehat{Z}$

The two constants a_1 and a_2 are determined from the conditions $\widehat{u}\left(\widehat{x}\right) = \widehat{u}_1$ at $\widehat{x} = 0$ and $\widehat{u}\left(\widehat{x}\right) = \widehat{u}_2$ at $\widehat{x} = l$. This gives

$$\begin{aligned} \widehat{u}\left(\widehat{x} = 0\right) &= \widehat{u}_1 = a_1 \\ \widehat{u}\left(\widehat{x} = l\right) &= \widehat{u}_2 = a_1 + a_2 l = \widehat{u}_1 + a_2 l \Rightarrow a_2 = \frac{\widehat{u}_2 - \widehat{u}_1}{l} \end{aligned} \tag{4.21}$$

Equation (4.20) then becomes

$$\begin{aligned} \widehat{u}\left(\widehat{x}\right) &= \widehat{u}_1 + \frac{\widehat{u}_2 - \widehat{u}_1}{l}\widehat{x} \\ &= \left(1 - \frac{\widehat{x}}{l}\right)\widehat{u}_1 + \frac{\widehat{x}}{l}\widehat{u}_2 \end{aligned} \tag{4.22}$$

The above shape function is written in matrix form given by

$$\begin{aligned} \widehat{u}\left(\widehat{x}\right) &= \left[1 - \frac{\widehat{x}}{l} \quad \frac{\widehat{x}}{l}\right]\left\{\begin{array}{c} \widehat{u}_1 \\ \widehat{u}_2 \end{array}\right\} \\ &= [N_1 \quad N_2]\left\{\begin{array}{c} \widehat{u}_1 \\ \widehat{u}_2 \end{array}\right\} \end{aligned} \tag{4.23}$$

The shape functions in the above are

$$\begin{aligned} N_1 &= 1 - \frac{\widehat{x}}{l} \\ N_2 &= \frac{\widehat{x}}{l} \end{aligned} \tag{4.24}$$

We also denote the nodal degrees of freedom by a vector

$$\widehat{d} = \left\{\begin{array}{c} \widehat{u}_1 \\ \widehat{u}_2 \end{array}\right\} \tag{4.26}$$

We can then write

$$\widehat{u} = [N]\left\{\widehat{d}\right\} \tag{4.27}$$

Step 3: Strain-Displacement Relations

In this case strain displacement relations are simple given by Eq. (4.2)

$$\varepsilon_{ij} = \frac{1}{2}\left(\frac{\partial u_i}{\partial x_j} + \frac{\partial u_j}{\partial x_i}\right) \Rightarrow \varepsilon_{xx} = u_{,x} \quad \text{rest } \varepsilon = 0$$
$$\therefore \varepsilon_x = \begin{bmatrix} -\frac{1}{l} & \frac{1}{l} \end{bmatrix}\left\{\hat{d}\right\} = \begin{bmatrix} N_{1,\hat{x}} & N_{2,\hat{x}} \end{bmatrix}\left\{\hat{d}\right\} = [B]\left\{\hat{d}\right\} \tag{4.28}$$

In the above $[B] = \begin{bmatrix} N_{1,\hat{x}} & N_{2,\hat{x}} \end{bmatrix}$ is Strain-Displacement matrix.

Step 4: Strain Energy

The strain energy in the element is given by (4.4)

$$U = \frac{1}{2}\int_0^l EA\hat{u}_{,x}^2\, d\hat{x} \tag{4.30}$$

Using (4.28) in the above

$$U = \frac{1}{2}EA\left\{\hat{d}\right\}^T \int_0^l [B]^T[B]\left\{\hat{d}\right\}d\hat{x} \tag{4.31}$$

Step 5: Elemental Stiffness Matrix

Now we apply Eq. (4.19)

$$\frac{\partial U}{\partial\left\{\hat{d}\right\}} = \frac{\partial}{\partial\left\{\hat{d}\right\}}\left[\frac{1}{2}EA\left\{\hat{d}\right\}^T \int_0^l [B]^T[B]\left\{\hat{d}\right\}d\hat{x}\right] = \{P\}$$

i.e.,

$$\left[EA\int_0^l [B]^T[B]\left\{\hat{d}\right\}d\hat{x}\right] = \{P\}$$
i.e.,
$$\left[\hat{K}\right]\left\{\hat{d}\right\} = \{P\} \tag{4.32}$$

The stiffness matrix in the above is

$$\left[\widehat{K}\right] = EA \int_0^l [B]^T [B] d\widehat{x} \tag{4.33}$$

Now substituting for $[B]$ from (4.28)

$$\begin{aligned} \left[\widehat{K}\right] &= EA \int_0^l \begin{bmatrix} N_{1,\widehat{x}} \\ N_{2,\widehat{x}} \end{bmatrix} \begin{bmatrix} N_{1,\widehat{x}} & N_{2,\widehat{x}} \end{bmatrix} d\widehat{x} \\ &= EA \int_0^l \begin{bmatrix} -\frac{1}{l} \\ \frac{1}{l} \end{bmatrix} \begin{bmatrix} -\frac{1}{l} & \frac{1}{l} \end{bmatrix} d\widehat{x} \end{aligned} \tag{4.34}$$

Performing the integration

$$\left[\widehat{K}\right] = \frac{EA}{l} \begin{bmatrix} 1 & -1 \\ -1 & 1 \end{bmatrix} \tag{4.35}$$

Notice that the stiffness matrix is symmetric and therefore we can write the diagonal and terms above the diagonal, the terms below the diagonal need not be written. In this case it does not matter since only one term is involved with $k_{12} = k_{21}$, however when we assemble several elements into a global matrix, it becomes easier.

$$[K] = \frac{EA}{l} \begin{bmatrix} 1 & -1 \\ & 1 \end{bmatrix} \tag{4.35a}$$

EA in the above represents the rigidity for axial deformation, *axial rigidity*.

4.3.1 Example of a Simple Column with an Axial Load

Let us consider the problem of a simple vertical load taken by a column in Fig. 2.10. Let this be treated as one element only by the finite element method. Then at left end in Fig. 4.6 (or at bottom end in Fig. 2.10) on node 1 the displacement $d_1 = 0$; with the axial load applied on the right end (or top end) node 2. Here the element 1 is the structure, therefore

$$\left[\widehat{K}\right]\left\{\widehat{d}\right\} = \{P\} \Rightarrow [K]\{d\} = \{P\} \tag{4.36}$$

Substituting

$$\frac{EA}{l}\begin{bmatrix} 1 & -1 \\ & 1 \end{bmatrix}\begin{Bmatrix} d_1 \\ d_2 \end{Bmatrix} = \begin{Bmatrix} 0 \\ P \end{Bmatrix} \tag{4.37}$$

The above simplifies to

$$\begin{gathered} \frac{EA}{l}\begin{bmatrix} 1 & -1 \\ & 1 \end{bmatrix}\begin{Bmatrix} 0 \\ d_2 \end{Bmatrix} = \begin{Bmatrix} 0 \\ P \end{Bmatrix} \\ \frac{EA}{l} d_2 = P \Rightarrow d_2 = \frac{Pl}{EA} \end{gathered} \tag{4.37a}$$

From (4.23) we have

$$u(x) = \begin{bmatrix} 1 - \frac{x}{l} & \frac{x}{l} \end{bmatrix}\begin{Bmatrix} 0 \\ u_2 \end{Bmatrix} = \frac{Px}{EA} \tag{4.38}$$

The strain in the column is

$$\varepsilon_x = u_{,x} = \frac{P}{EA} \tag{4.39}$$

The stress is

$$\sigma_x = E\varepsilon_x = \frac{P}{A} \tag{4.40}$$

Note that both the strain and stress are constant throughout the column.

4.3.2 *Example of a Two Stepped Column with an Axial Load*

Let us now consider a more complicated structure in Fig. 4.7 that can be easily solved by the finite element method discussed above.

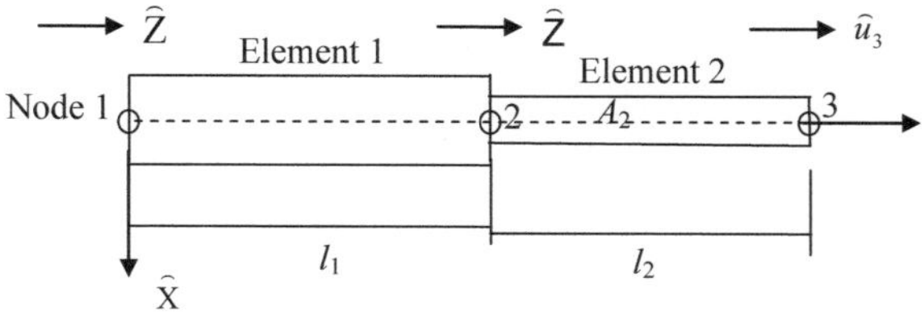

Fig. 4.7 Two stepped bar

We have now three nodes and two elements, element 1 with nodes 1 and 2 with area A_1, length l_1 and element 2 with nodes 2 and 3 with area A_2, length l_2. The load P is applied at node 3. For element 1,

$$\begin{aligned}\left[\widehat{K}_1\right]\left\{\widehat{d}_1\right\} &= \{0\} \\ \frac{EA_1}{l_1}\begin{bmatrix} 1 & -1 \\ & 1 \end{bmatrix}\begin{Bmatrix} \widehat{u}_1 \\ \widehat{u}_2 \end{Bmatrix} &= \begin{Bmatrix} 0 \\ 0 \end{Bmatrix}\end{aligned} \tag{4.41}$$

For element 2,

$$\left[\widehat{K}_2\right]\left\{\widehat{d}_2\right\} = \frac{EA_2}{l_2}\begin{bmatrix} 1 & -1 \\ & 1 \end{bmatrix}\begin{Bmatrix} \widehat{u}_2 \\ \widehat{u}_3 \end{Bmatrix} = \begin{Bmatrix} 0 \\ P \end{Bmatrix} \tag{4.42}$$

For the bar we now assemble the elemental stiffness matrices to get the global stiffness matrix.

$$\begin{bmatrix} \frac{EA_1}{l_1} & -\frac{EA_1}{l_1} & \\ & \frac{EA_1}{l_1} + \frac{EA_2}{l_2} & -\frac{EA_2}{l_2} \\ & & \frac{EA_2}{l_2} \end{bmatrix}\begin{Bmatrix} u_1 = 0 \\ u_2 \\ u_3 \end{Bmatrix} = \begin{Bmatrix} 0 \\ 0 \\ P \end{Bmatrix} \tag{4.43}$$

Taking away the column and row of node 1

$$\begin{bmatrix} \frac{EA_1}{l_1} + \frac{EA_2}{l_2} & -\frac{EA_2}{l_2} \\ & \frac{EA_2}{l_2} \end{bmatrix}\begin{Bmatrix} u_2 \\ u_3 \end{Bmatrix} = \begin{Bmatrix} 0 \\ P \end{Bmatrix} \tag{4.44}$$

$$\begin{aligned}\left(\frac{EA_1}{l_1} + \frac{EA_2}{l_2}\right)u_2 - \frac{EA_2}{l_2}u_3 &= 0 \\ -\frac{EA_2}{l_2}u_2 + \frac{EA_2}{l_2}u_3 &= P\end{aligned} \tag{4.44a}$$

First determine u_2 in terms of u_3

$$u_2 = \frac{\frac{EA_2}{l_2}}{\frac{EA_1}{l_1} + \frac{EA_2}{l_2}}u_3$$

Eliminate u_2

$$\left(-\frac{EA_2}{l_2}\frac{\frac{EA_2}{l_2}}{\frac{EA_1}{l_1} + \frac{EA_2}{l_2}} + \frac{EA_2}{l_2}\right)u_3 = P$$

The deflection at the right end is

$$u_3 = P\frac{\frac{EA_1}{l_1} + \frac{EA_2}{l_2}}{\frac{EA_1}{l_1}\frac{EA_2}{l_2}} \tag{4.45}$$

If both the elements have area A and length 1/2 l the above gives $u_3 = \frac{Pl}{EA}$ as in (4.37a).

4.3.3 Simply Supported—Free Bar—One Element

In Sect. 4.2 we considered a bar simply supported at left end and free at right end with a distributed axial load $P(x)$ instead of nodal loads P in the earlier examples of 4.3.1 and 4.3.2. In a finite element method, we lump the distributed load in the elemental length at the left and right nodes, each equal to 1/2 Pl. This makes Eq. (4.37a)

$$\begin{gathered}\frac{EA}{l}\begin{bmatrix} 1 & -1 \\ & 1 \end{bmatrix}\begin{Bmatrix} 0 \\ d_2 \end{Bmatrix} = \begin{Bmatrix} 0 \\ \frac{1}{2}Pl \end{Bmatrix} \\ \frac{EA}{l}d_2 = \frac{1}{2}Pl \Rightarrow d_2 = \frac{1}{2}\frac{Pl^2}{EA}\end{gathered} \tag{4.37b}$$

The right end deflections then are same with lumped masses used in finite element method. Notice that the finite element method uses a linear interpolation function and between nodes 1 and 2 the displacement is linear, whereas in Lagrangian formulation it is parabolic. In both the case the nodal deflections are same, in finite element case it is a linear variation and Lagrange formulation it is parabolic. This is explained in Fig. 4.5—as we increase the number of elements, i.e., decrease the elemental length we get converged results.

4.3.4 Simply Supported—Free Bar—Two Elements

With the two elements identical of length 1/2 l, Eq. (4.43) can be written as

$$\frac{2EA}{l}\begin{bmatrix} 1 & -1 & \\ & 2 & -1 \\ & & 1 \end{bmatrix}\begin{Bmatrix} u_1 = 0 \\ u_2 \\ u_3 \end{Bmatrix} = \begin{Bmatrix} \frac{1}{4}Pl \\ \frac{1}{2}Pl \\ \frac{1}{4}Pl \end{Bmatrix} \tag{4.43a}$$

$$\frac{2EA}{l}\begin{bmatrix} 2 & -1 \\ & 1 \end{bmatrix}\begin{Bmatrix} u_2 \\ u_3 \end{Bmatrix} = \begin{Bmatrix} \frac{1}{2}Pl \\ \frac{1}{4}Pl \end{Bmatrix} \tag{4.44a}$$

$$\frac{2EA}{l}(2u_2 - u_3) = \frac{1}{2}Pl$$

$$\frac{2EA}{l}(-u_2 + u_3) = \frac{1}{4}Pl \tag{4.37b}$$

$$\Rightarrow u_3 = \frac{Pl^2}{2EA}$$

The one element and two element models give the same free end results with linear variation between the nodes and agree with the Lagrangian formulation at the free end which gives parabolic variation.

We see the main advantage of the finite element method that we can discretize a given structure into several elements of uniform properties, whose stiffnesses in their own elemental coordinates are first obtained and assembled to give the global stiffness matrix. The nodal loads are determined by lumping the distributed loads if they are there and when the elemental size decreases these loads will approximate to the distributed load. Applying the boundary conditions we can then obtain a set of linear equations that are solved to determine the deflections, strains and stresses at each node point. If one wants to do this computation by hand it becomes tedious and not possible when the number of elements is more than three or four. In the early days of computers, one wrote their own computer codes; today because of extensive usage of these methods, commercial codes are developed which are rugged even for a million or more elements that can easily handle any complex structure. The time consuming hand calculations are now practically replaced just as slide rules replaced log tables before 1960s and calculators replaced slide rules in 1980s. Today one depends on computers to solve many engineering problems.

Some of the examples given below use a suitable commercial code (there are many available) to the faculty or student. The procedures or training on these codes is left to individual student or faculty. The solutions should invariably be checked using simple cases and closed form solutions or with few elements by hand calculations. The advantage of practicing this line of solving the problems is to get into practice of handling complex geometry structures with large number of elements which cannot be done hand calculations.

Worked Example 4.1

Consider a circular cross-section bar as shown in Fig. Worked Example 4.1a. It is clamped on the left side and has a point force of 10 kN acting axially on the right end of the beam. The bar is 1 m long with a radius 0.05 m; additionally, the bar is made of steel which has a Young's Modulus 200 GPa.

Schematically the one dimensional equivalent structure is given in Fig. Worked Example 4.1b.

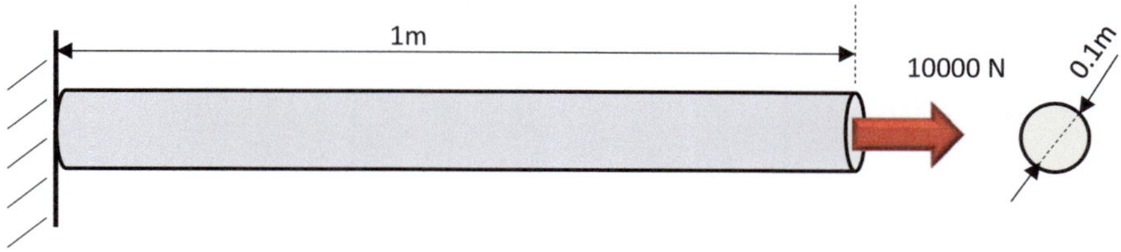

Fig. Worked Example 4.1a Axially loaded bar

Fig. Worked Example 4.1b Axially loaded bar

The maximum deformation of the bar is calculated using

$$u = \frac{Pl}{EA} = 0.006366198 \text{ mm}$$

The maximum axial stress is calculated using $\sigma = \frac{P}{A} = 1.273239545$ MPa

One can use a given commercial code for this purpose. Modeling it as one element, the solution obtained is shown in Fig. Worked Example 4.1c.

The solution from a FEM 0.006377116 mm and closed form solution 0.006366198 mm are in good agreement. The FEM predicts an upper bound value. Notice that the commercial codes give a color coded display identifying the deformation in various locations. Here nine ranges are chosen.

Equation (4.10) gives the deformation solution which is parabolic; however the shape function adopted in Eq. (4.20) is linear. One can adopt a shape function to be of higher order, and then the solution can match closely with the exact solution. If you choose such an element from commercial code library we will expect better results even with fewer elements. Here a cubic interpolation function element is adopted.

If you have chosen to use 2 elements for your mesh, you should see the following output (Fig. Worked Example 4.1d).

Fig. Worked Example 4.1c Axially loaded bar—one element solution for deformation mm

Fig. Worked Example 4.1d Axially loaded bar—two element solution for deformation mm

Notice the two elements and nine color bands for the deformation. Maximum deformation remains same as before.

If you have chosen to use 10 elements for your mesh, you should see the following output for the total deformation (Fig. Worked Example 4.1e).

You can see the ten elements and nine color bands for the deformation. The maximum deformation has not changed; this is because the assumed shape function is linear.

The axial stress obtained with 10 elements is shown in Fig. Worked Example 4.1f.

The stress obtained is constant at 1.2742 MPa; the force acting at any cross-section being same in axially loaded bar, the stress is also same. The analytical calculation for stress is 1.27324 MPa. The FEM result is an upper bound value.

Notice that the stress value with one element is larger and with increase in number of elements used, it converges rapidly to a lower value, closer to the analytical calculation.

Table Worked Example 4.1 gives the FEM outputs for a mesh of 1, 2 elements and a mesh of 10 elements. One is advised to perform a convergence study to make sure that the solution obtained is correct and that no mistakes have occurred in modeling and meshing.

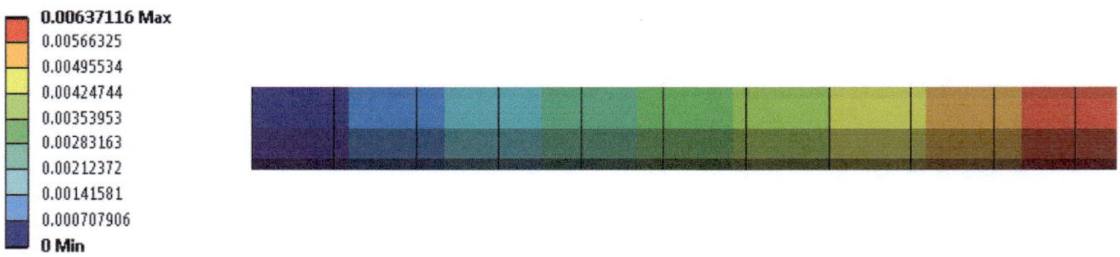

Fig. Worked Example 4.1e Axially loaded bar—ten element solution for deformation mm

Fig. Worked Example 4.1f Axially loaded bar—ten element solution for stress MPa

Table Worked Example 4.1 Convergence study with increasing number of elements

	Total deformation (m)	Axial stress (MPa)
Theory values	0.006366198	1.273239545
1 element FEA	0.00637116	1.2742
2 element FEA	0.00637116	1.2739
10 element FEA	0.00637116	1.2739

As the number of elements increase, the stress decreases and converges to a slightly higher value than the exact value.

Worked Example 4.2

Consider a two-stepped bar shown in Fig. Worked Example 4.2a. Use $E = 200$ GPa.

The deflection at the right end from Eq. (4.45) is

$$u_3 = P\frac{\frac{EA_1}{l_1} + \frac{EA_2}{l_2}}{\frac{EA_1}{l_1}\frac{EA_2}{l_2}} = 0.01591549 \text{ mm}$$

The maximum bending stress with small cross-section of diameter 0.05 m is

$$\sigma = \frac{P}{A} = 5.092958179 \quad \text{MPa}$$

An FE model of the two stepped bar in Fig. Worked Example 4.2a is made and analyzed in a commercial code. The total deformation is given in Fig. Worked Example 4.2b and the axial stress result is given in Fig. Worked Example 4.2c.

The FE result is 0.0159279 mm as compared to the analytical result 0.01591549 mm. As in the previous example, the FE result is upper bound. The results are compared in Table Worked Example 4.2.

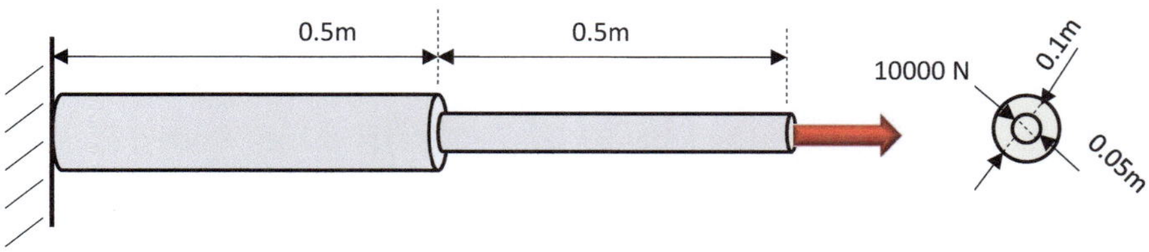

Fig. Worked Example 4.2a Two stepped bar

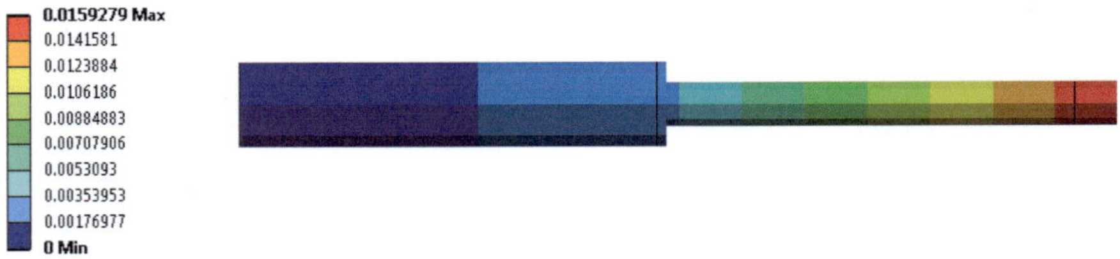

Fig. Worked Example 4.2b Two stepped bar—deformation mm

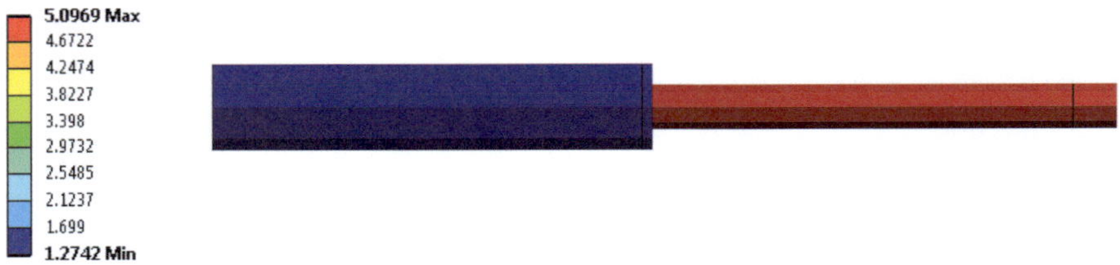

Fig. Worked Example 4.2c Two stepped bar—axial stress

Table Worked Example 4.2 Comparisons of analytical and FEA results

	Total deformation (mm)	Axial stress (MPa)
Theory values	0.01591549	5.092958179
1 element FEA	0.0159278	5.096921
2 element FEA	0.0159271	5.096124
10 element FEA	0.0159271	5.096124

Worked Example 4.3

A simply supported—free bar under an axial load at free end is given in Fig. Worked Example 4.3a. Use E = 200 GPa.

The analytical result for deflection at the right end from (4.15) is

$$u = \frac{Pl^2}{EA} = 0.0063662 \text{ mm}$$

The model with one, two and ten elements is shown in Fig. Worked Example 4.3b.

The FE Result for one element is shown in Fig. Worked Example 4.3c.

The results obtained are compared in Table Worked Example 4.3.

The FE solution is once again upper bound. The solution converges from 1 element solution, down from 0.00642032 to 0.00637618 mm, which is slightly above the theoretical value.

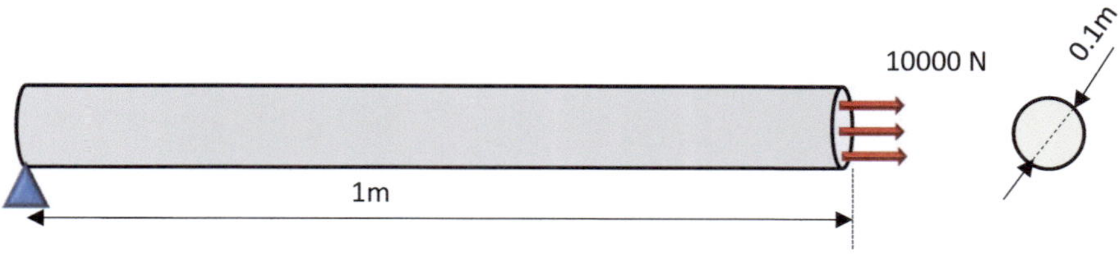

Fig. Worked Example 4.3a Bar under axial load at free end

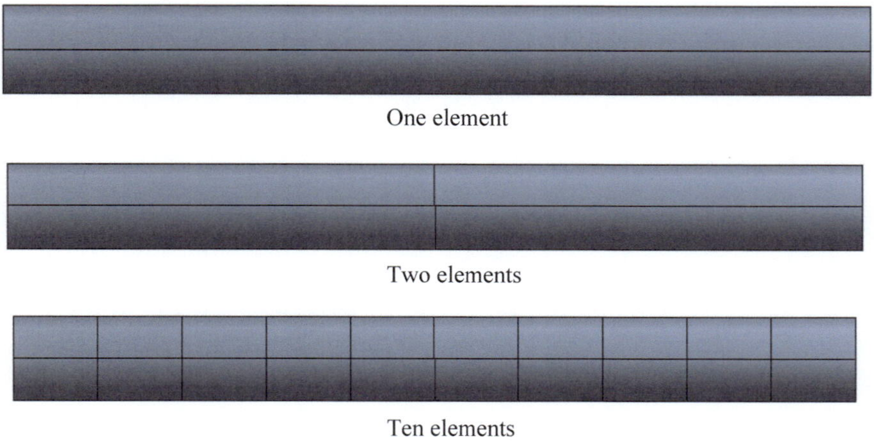

Fig. Worked Example 4.3b FEA model, number of elements 1, 2 and 10

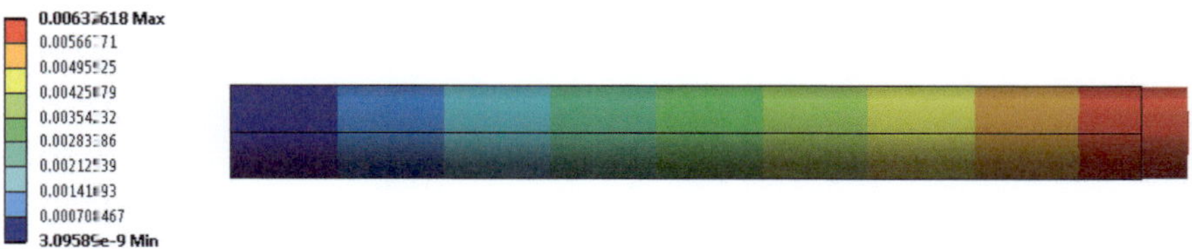

Fig. Worked Example 4.3c FE result mm for one element

Table Worked Example 4.3 Comparison of results

	Total deformation (mm)
Theory	0.0063662
1 element FEA	0.00642032
2 element FEA	0.00644023
10 element FEA	0.00637618

From Sect. 4.2 the stress at the supported end $\frac{Pl}{A} = \frac{10{,}000}{\pi\times\frac{0.1^2}{4}} = 1.273$ MPa and drops linearly to zero at the free end.

4.4 Thermal Stresses

Many engineering components are subjected to temperature changes that result in a change in length. An elongation thus obtained causes thermal strain. If this elongation is allowed to take place freely, then there is no stress associated with the thermal strain unless the elongation is restrained by the supports.

The additional support can be taken as redundant and in its place we can apply the principle of superposition. Let the length of a bar subjected to thermal expansion be L as shown in Fig. 4.8. Let a be the thermal coefficient of expansion.

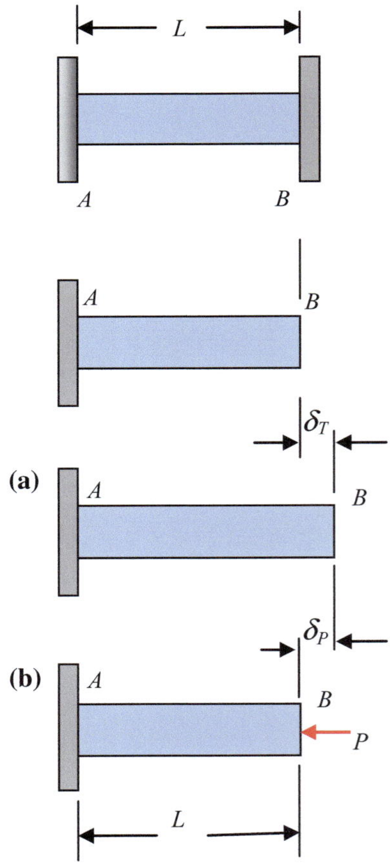

Fig. 4.8 A bar under thermal expansion

If the right support is made free as in Fig. 4.8a, the thermal expansion δ_T is given by

$$\delta_T = \alpha \times \Delta T \times L \tag{4.46}$$

The right end of the bar moves by the above distance in (4.46) as seen in Fig. 4.8b. If we restrict the expansion with the right support at B, an equivalent load compatible to this condition P is applied. This equivalent load is

$$\delta_P = \frac{PL}{AE} \tag{4.47}$$

The thermal deformation and the deformation from the redundant support must be compatible, i.e.,

$$\delta = \delta_T + \delta_P = 0 \tag{4.48}$$

Therefore

$$\begin{aligned} \delta &= \alpha \times \Delta T \times L + \frac{PL}{AE} = 0 \\ \Rightarrow P &= -AE\alpha \times \Delta T \\ \Rightarrow \sigma &= \frac{P}{A} = -E\alpha \times \Delta T \end{aligned} \tag{4.49}$$

We therefore see that a thermal expansion problem is equivalent to an axially loaded bar with an equivalent nodal load P defined above.

Worked Example 4.4

A steam pipe is 30 m long at a temperature of 15 °C. When steam at 180 °C is made to enter how much expansion of the pipe takes place? If you prevent the expansion what stress and strain will be induced in the pipe material? The thermal coefficient of expansion α is 0.000012 per °C and the elastic modulus is 200 GPa.

From (4.46)

$$\begin{aligned} \delta_T &= \alpha \times \Delta T \times L \\ &= 0.000012 \times 165 \times 30 = 0.0594\ \text{m} \end{aligned}$$

Strain due to prevention of expansion $\frac{0.0594}{30} = 0.00198$

Stress from (4.49) is $\sigma = E\alpha \times \Delta T = 200{,}000 \times 0.000012 \times 165 = 396$ MPa

4.5 Principle of Virtual Work

Virtual work on a particle is defined as the work done by all the forces acting on it for a given small hypothetical displacement, called *virtual displacement*, consistent with any constraints present. The applied forces are all kept constant during this hypothetical displacement.

We can apply this principle to find the deflection of pin A in Fig. 4.9 from the force P inclined at an angle β. There are four bars, each of them of length L_i inclined at an angle α_i with the horizontal as shown. We note to begin with that the forces in the bars cannot be determined by equilibrium approach treating the bars as rigid. Such of these structures are *Statically Indeterminate structures*. Principle of virtual work becomes useful in such cases.

Let us now determine the forces in the bars for linearly elastic behavior using the Principle of Virtual Work.

The movement of pin A defines the deformation of the system. The only pin that is not constrained is pin A. Under the force P let it suffer displacement components $\bar{u}$ and $\bar{v}$ in X and Y directions respectively. The strain $\bar{\varepsilon}_i$ in the ith bar from these displacement components is

$$\bar{\varepsilon}_i = \frac{\bar{u}\cos\alpha_i}{L_i} + \frac{\bar{v}\sin\alpha_i}{L_i} \tag{4.50}$$

The corresponding stress components are

$$\bar{\sigma}_i = E\frac{\bar{u}\cos\alpha_i}{L_i} + E\frac{\bar{v}\sin\alpha_i}{L_i} \tag{4.51}$$

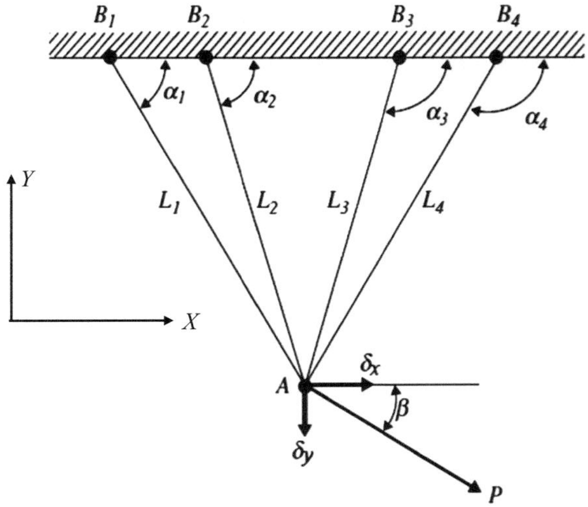

Fig. 4.9 Four constrained bars pinned at A

Let us impose a virtual displacement δ_x in the X direction that causes the bars change in length and rotate about the fixed ends B_i without violating any constraints. Then each bar i has a virtual change in length L_i given by

$$\delta L_i = \delta_x \cos \alpha_i \tag{4.52}$$

Therefore the virtual strain in each bar i is

$$\delta \varepsilon_i = \frac{\delta_x \cos \alpha_i}{L_i} \tag{4.53}$$

We have now a system of bars under the external force P subjected to strains and stresses according to (4.50) and (4.51) subjected to the virtual displacement and strain field given by (4.52) and (4.53). Now we use the principle of virtual work. The virtual work of the applied force P is equal to the virtual work done by virtual deformation on the stress field in all the four bars.

$$P \cos \beta \times \delta_x = \sum_{i=1}^{4} \bar{\sigma}_i A_i \times \delta \varepsilon_i L_i \tag{4.54}$$

Using (4.51) and (4.53), cancelling δ_x and simplifying, we have

$$P \cos \beta \times \delta_x \Longrightarrow \sum_{i=1}^{4} \left\{ E \frac{\bar{u} \cos \alpha_i}{L_i} \left(\frac{\delta_x \cos \alpha_i}{L_i} \right) + E \frac{\bar{v} \sin \alpha_i}{L_i} \left(\frac{\delta_x \cos \alpha_i}{L_i} \right) \right\} A_i L_i$$

$\Rightarrow$

$$P \cos \beta = \sum_{i=1}^{4} \left(\bar{u} \cos^2 \alpha_i + \bar{v} \sin \alpha_i \cos \alpha_i \right) \frac{EA_i}{L_i} \tag{4.55}$$

Now we give Pin A the second virtual displacement δ_y, then

$$P \sin \beta = \sum_{i=1}^{4} \left(\bar{u} \sin \alpha_i \cos \alpha_i + \bar{v} \sin^2 \alpha_i \right) \frac{EA_i}{L_i} \tag{4.56}$$

We can now solve the above two Eqs. (4.55) and (4.56) simultaneously for the two unknowns $\bar{u}$ and $\bar{v}$. The actual stresses and forces in the bars can then be readily computed.

The virtual displacement is also called the *dummy displacement* and the Principle of Virtual work method *Dummy Displacement* method.

Virtual Work Principle using Delta Operator:

Notice that we conveniently denoted a virtual displacement by employing the Variational operator δ. Thus, δu_i represents a virtual displacement field for a given configuration u_i.

We can generalize the Virtual Work Principle using the delta operator notation as

$$\iiint_V B_i \delta u_i dV + \iint_S T_i^{(v)} \delta u_i dS = \iiint_V \sigma_{ij} \delta \varepsilon_i dV \tag{4.57}$$

The left side is the external virtual work of the body forces B_i and traction forces $T_i^{(v)}$ on the surface S and the right side is the internal virtual work. We can thus use the principle of virtual work directly in any statically indeterminate structural problem where we can completely specify a virtual displacement field by giving the variation of a finite number of parameters. Thus for trusses, where we have axially loaded members, we need to only specify the virtual displacements of the pins or joints.

Complementary Virtual Work:

Thus far, we varied the displacement field while keeping the external forces fixed. Instead we can vary the stress field and external forces while holding displacement fields fixed. The variations of stress and external loads will be allowed only those which satisfy the equations of equilibrium inside the body and on the boundary. That is

$$\begin{aligned} \left(\delta\sigma_{ij}\right)_{,j} + \delta B_i &= 0 \quad \text{in } V \\ \left(\delta\sigma_{ij}\right) v_j = \delta T_i^{(v)} &= 0 \quad \text{on } S_2 \end{aligned} \tag{4.58}$$

On S_1 boundary it is assumed that $T_i^{(v)}$ is specified, therefore $\delta T_i^{(v)} = 0$ there. We can therefore extend S_2 to S. The *complementary virtual work* δW^* is defined as

$$\delta W^* = \iiint_V u_i \delta B_i dV + \iint_S u_i \delta T_i^{(v)} dS \tag{4.59}$$

We can expand the above

$$\begin{aligned} \delta W^* &= \iiint_V u_i \delta B_i dV + \iint_S u_i \delta T_i^{(v)} dS \\ &= \iiint_V u_i \delta B_i dV + \iint_S u_i \left(\delta\sigma_{ij}\right) v_j dS \\ &= \iiint_V u_i \delta B_i dV + \iiint_V \left(u_i \delta\sigma_{ij}\right)_{,j} dV \\ &= \iiint_V u_i \left[\delta B_i + \left(u_i \delta\sigma_{ij}\right)_{,j}\right] dV + \iiint_V u_{i,j} \delta\sigma_{ij} dV \end{aligned} \tag{4.60}$$

Dropping the first term on the right-hand side and replacing the deformation field by strain field, we have

$$\iiint_V u_i \delta B_i dV + \iint_S u_i \delta T_i^{(v)} dS = \iiint_V \varepsilon_{ij} \delta\sigma_{ij} dV \tag{4.61}$$

The principle of complementary virtual work can be used directly for solving certain structural problems. In this case we can very easily find deflections of any particular joint in any particular direction for statically determinate structures. Statically determinate structures allow us to find the forces taken by each member under a given loading conditions by using a rigid body equilibrium approach. That leads us to the stress and strain field, in the above Eq. (4.61) to the strain field. Figure 4.9 is on the other hand a statically indeterminate structure to determine the forces in each member by rigid body equilibrium. There virtual work principle with virtual deflection field is used to get the solution of the displacements first.

Consider a plane truss consisting of 9 members (bars) supported at pins A and B as in Fig. 4.10. There are 4 pins or joints marked in red circles other than at the supports. Forces P_1 and P_2 act vertically down at the two pins 1 and 2.

Here we first perform an equilibrium analysis treating the members as rigid under the actual loads P_1 and P_2 and determine the reactions A_x and A_y at the support A and the vertical reaction B_y at the support B and thus the forces F_i in each member. This is not described here; it leads to the actual stress and strain field ε_i.

$$\varepsilon_i = \frac{F_i}{EA_i} \tag{4.62}$$

Noting that there are no body forces, we can use Eq. (4.61) to determine u_α along $\hat{n}$ under a unit force as follows. The traction loads (forces) and stresses are varied.

The left side of Eq. (4.61) becomes simply u_α since $\delta T_i^{(v)}$ is just the unit force at pin 4. Now going to the right side of the equation, we by first compute the virtual forces

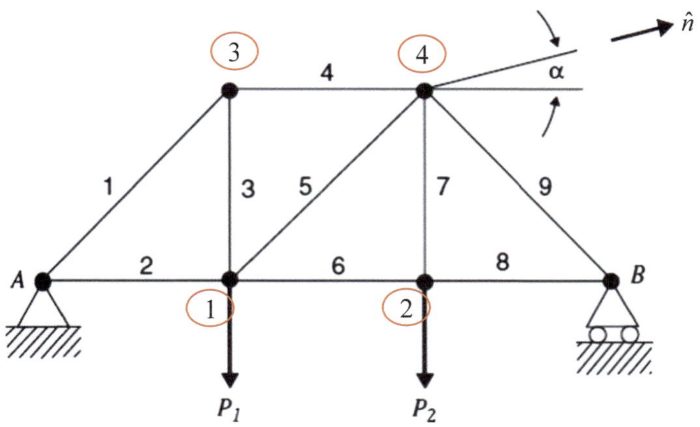

Fig. 4.10 Plane truss with two forces

$(F_v)_i$ in each member from the unit force at pin 4 using equilibrium considerations of rigid-body mechanics. Using these virtual forces we determine the varied stresses in each member $\delta\sigma_{ij} = \frac{(F_V)_i}{A_i}\delta\sigma_{ij}$ are determined. Then Eq. (4.61) yields

$$\iint\limits_S u_i \delta T_i^{(v)} dS = \iiint\limits_V \varepsilon_{ij}\delta\sigma_{ij} dV$$

$$\Rightarrow u_\alpha = \sum_{i=1}^{8} \int_0^{L_i} \frac{F_i}{EA_i}\left\{\frac{(F_V)_i}{A_i}\right\} A_i dx \qquad (4.63)$$

$$= \sum_{i=1}^{8} \frac{F_i (F_V)_i}{EA_i} L_i$$

4.6 Minimization of Total Potential Energy

Consider the simple column of Sect. 4.3.1 we assume the deflection or shape function to be

$$u(x) = ax$$
$$\frac{du}{dx} = u_{,x} = a \qquad (4.64)$$

The total potential energy π is

$$\pi = U - W = \int_0^l \left(\frac{1}{2} EAu_{,x}^2 - Pu\right) dx$$
$$= \int_0^l \left(\frac{1}{2} EAa^2\right) dx - Pu_{x=l} \qquad (4.65)$$
$$= \frac{1}{2} EAla^2 - Pal$$

Minimize the total potential energy by making its derivative with respect to a to be zero

$$\frac{\partial \pi}{\partial a} = aEAl - Pl = 0 \Rightarrow a = \frac{P}{EA} \qquad (4.66)$$

The deflection $u(x)$ is then given by (4.38), and the strain and stress by (4.39) and (4.40)

$$u(x) = \frac{P}{EA}x \tag{4.38a}$$

The tip elongation is then

$$u_{\text{tip}} = \frac{Pl}{EA}$$

The principle of minimization of energy to determine the arbitrary constants in shape functions makes the calculation procedure simple. This principle is named after Walther Ritz (1878–1909). This principle makes it easy for structures which have variable cross-sectional areas.

4.6.1 Tapered Bar

Let the area of cross-section of the bar vary from A_0 to $A_1 = 0.5A_0$ in a linear manner along the length of the bar from 0 to l. Let this linear variation be $A = A_0\left(1 - \alpha\frac{x}{l}\right)$ where α is a taper parameter. This parameter $\alpha = 0.5$ for the present case. Equation (4.65) now becomes

$$\pi = U - W = \int_0^l \left(\frac{1}{2}EA_0\left(1 - 0.5\frac{x}{l}\right)u_{,x}^2\right)dx - Pu_{x=l} \tag{4.67}$$

For the shape function, we continue with the assumed one in Eq. (4.64). Therefore

$$\begin{aligned}\pi = U - W &= \int_0^l \left(\frac{1}{2}EA_0\left(1 - 0.5\frac{x}{l}\right)u_{,x}^2\right)dx - Pu_{x=l}\\ &= \int_0^l \left(\frac{1}{2}EA_0\left(1 - 0.5\frac{x}{l}\right)a^2\right)dx - Pal\\ &= \int_0^l \left(\frac{1}{2}EA_0a^2\right)dx - \int_0^l \left(\frac{1}{4}\frac{EA_0}{l}xa^2\right)dx - Pal\\ &= \frac{1}{2}EA_0la^2 - \frac{EA_0}{8}la^2 - Pal = \frac{3EA_0}{8}la^2 - Pal\end{aligned} \tag{4.68}$$

Now according to the Ritz principle

$$\frac{\partial \pi}{\partial a} = \frac{3}{4} a E A_0 l - Pl = 0 \Rightarrow a = \frac{4P}{3EA_0}$$

This gives

$$\begin{aligned} u &= \frac{4P}{3EA_0} x \\ u_{\text{tip}} &= \frac{4Pl}{3EA_0} \end{aligned} \tag{4.69}$$

With the bar tapered to a smaller area at the tip we have more elongation than before. The stain and stress in the bar are given by

$$\begin{aligned} \varepsilon_x = u_{,x} &= \frac{4P}{3EA_0} \\ \sigma_x = E\varepsilon_x &= \frac{4P}{3A_0} \end{aligned} \tag{4.70}$$

Both the strain and stress are constant irrespective of the taper in the cross-section.

4. Exercises

4.1. For slender bars justify that the strain field can be simplified to a single component and therefore that the stress is also simplified to a single component in the direction of the load.
4.2. Outline a procedure for determining the functional of a slender bar in axial load to be used in Euler-Lagrange equation.
4.3. For a slender bar, derive from Lagrange equation the governing differential equation and the boundary conditions.
4.4. A metallic bar 25 mm diameter is subjected to a tension test under a pull of 40 kN. The total extension determined over a length 20 cm is 0.08 mm. Determine modulus of elasticity of the material.
4.5. The maximum compression pressure in an Internal Combustion engine is 4 MPa acting on a piston whose diameter is 300 mm. The cylinder is held by four bolts whose effective diameter is 5 cm and length is 90 cm. The young's modulus of bolt material is 200 GPa. Determine maximum tensile stress.
4.6. Two thick aluminum alloy plates are held together in contact by copper bolts. The temperature is raised from 15 to 30 °C. Determine the increase in tensile

stress. The modulus of elasticity of copper is 117 GPa. Linear thermal expansion coefficients of aluminum alloy and copper respectively 0.000022 and 0.0000185 per °C.

4.7. Three bars each 300 mm^2 in cross-section and 2 m long support a force of 20,000 N. The center rod is made of steel and the two outer rods are made of copper. The temperature of all the rods is increased by 100 °C and made to elongate equally, determine the loads carried by each rod. Young's modulus of steel is 200 GPa and that of copper is 117 GPa; Linear thermal expansion coefficients of steel and copper respectively 0.000012 and 0.0000185 per °C.

4.8. Explain the concept of finite element method and show how you assume the shape function for axially loaded bars.

4.9. Discuss the energy principle. What is total potential energy of an axially loaded bar. Using the principle that Variation of the total potential energy is zero for a structure derive the stiffness matrix of an axially loaded bar.

4.10. Explain from fundamentals, the method of assembling elemental stiffness matrices. Take a three step bar and give the assembled stiffness matrix of an axially loaded bar.

4.11. Determine the forces carried by all the members shown in Fig. Exercises 4.11 with the help of the method of virtual work. Also determine the horizontal movement of pin *A*? The area of cross-section of all members is *a* modulus of elasticity *E*.

4.12. A load of 100 kg is carried by three axially loaded bars each of length 3 m, refer to Fig. Exercises 4.11. The bars are all of equal area of cross-section 12.5 mm^2 and made of steel with Young's modulus 200 GPa. Use the method of virtual work to find the forces in the members shown and determine the horizontal movement of pin A?

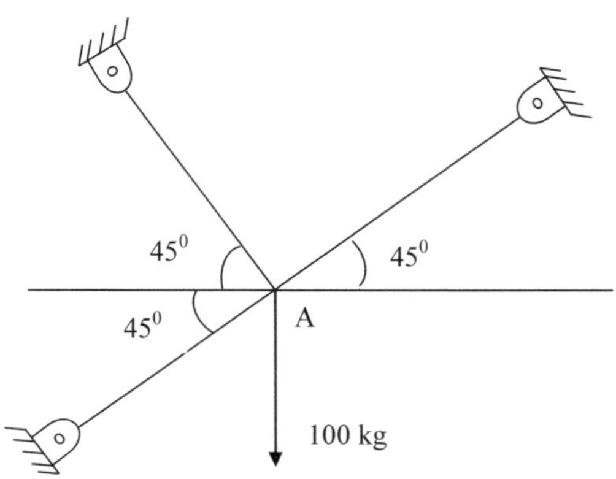

Fig. Exercises 4.11 Virtual Work Principle Exercise

4.13. Use the dummy load method to compute the vertical movement of joint A of the truss shown below. All the members are made of steel with a cross-sectional area equal to 15 mm^2.

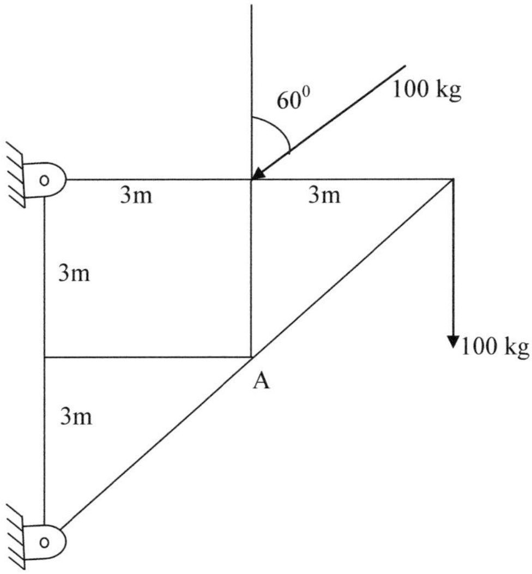

4.14. Using virtual work principle, determine the forces and stresses in all the members made of steel with an area of cross-section 15 mm^2.

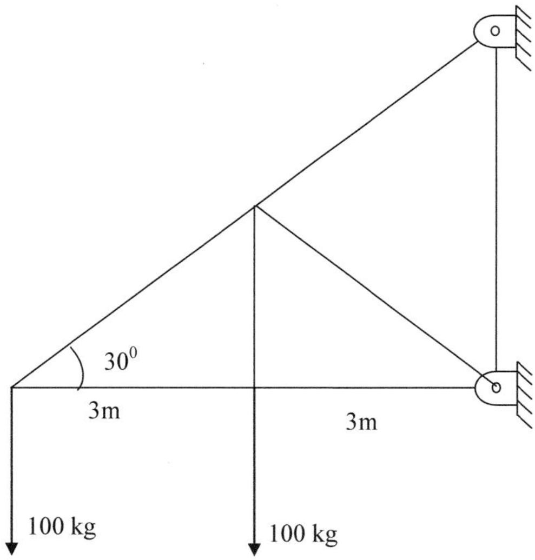

4.15. An axially loaded steel bar 2 m long and cross-sectional area 20 mm^2 is subjected to a 1000 N force on the right side. Determine the total elongation, strain and stress in the bar.

4.16. If the bar in the problem above is tapered from left to right from 20 mm^2 to half of this value, determine the increase in the total elongation. Also determine the percentage changes in strain and stress values. Use (a) Ritz method and (b) finite element method with a two stepped bar formulation of equal lengths.

Chapter 5
Twisting of a Rod

Another common structural member is a rod that is in twisting condition described by an angular deflection, also called torsion of a rod. In all rotating machines, the power is transmitted by shafts; the crankshaft of an automobile engine that transfers the motion through a clutch and gear box. The gear box has several shafts that carry the torque of the engine and then through a propeller shaft transmitting the torque to rear axles … All the shafts are torsion members suffering angular twist and shear strains and shear stresses. Just as 1D-Axial structures, the shafts transmitting the torque can also be treated as one-dimensional members, i.e., we recognize the rod by one coordinate, in this case the axis along the center line of the shaft. The twist θ is expressed as a function of x alone that keeps the theory to be simple and yet gives good results because the cross-sectional dimension of these shafts are very small compared to the length.

Consider a simple circular rod of uniform cross-section of area A shown in Fig. 5.1.

The rod is considered fixed at left end and subjected to a moment M applied at right end at $x = l$. We would like to know the twist $\theta\ (x)$. As we looked at axially loaded bar, we will keep the theory simple by assuming Poisson displacements are negligible so that we need only one independent variable x. The cross-section is circular that allows us to assume any out of plane displacements at any section x. If the section is non-circular, e.g., a rectangular cross-section, this section distorts with out-of-plane displacements. We will consider only circular section rods.

As in the previous Chap. 4 for axially loaded bars, we will follow the steps, beginning with an assumed displacement field. In the cross-section at x shown in Fig. 5.1 consider a point P(y, z). Considering the counter clockwise twist of this section, point P moves to P_1 through angle $\theta(x)$. Since no warping or out-of-plane displacement is assumed, we have

J.S. Rao, *Simulation Based Engineering in Solid Mechanics*,
DOI 10.1007/978-3-319-47614-8_5

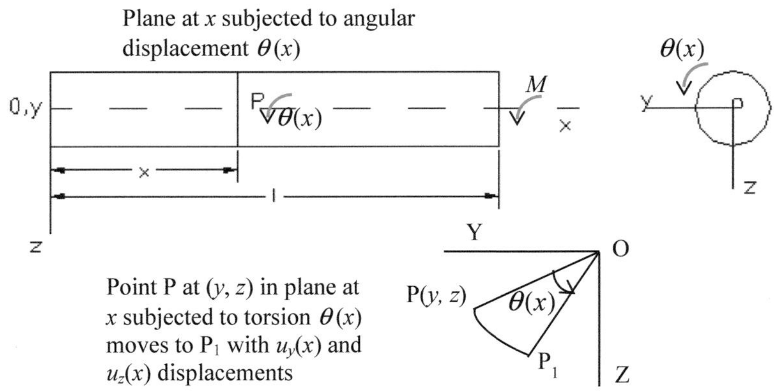

Fig. 5.1 A simple rod in torsion

1. Displacement field of a point P at coordinates x, y, z

$$\begin{aligned} u_x &= 0 \\ u_y &= -z\theta(x) \\ u_z &= y\theta(x) \end{aligned} \tag{5.1}$$

Let us introduce the rate of twist α to be linear along the axis x, i.e., $\alpha = \frac{d\theta}{dx}$. Then we have $\theta = \alpha x$ and (5.1) is rewritten as

$$\begin{aligned} u_x &= 0 \\ u_y &= -z\alpha x \\ u_z &= y\alpha x \end{aligned} \tag{5.1a}$$

2. Strain Field

$$\begin{aligned} &\begin{bmatrix} \varepsilon_{xx} & \varepsilon_{xy} & \varepsilon_{xz} \\ \varepsilon_{yx} & \varepsilon_{yy} & \varepsilon_{yz} \\ \varepsilon_{zx} & \varepsilon_{zy} & \varepsilon_{zz} \end{bmatrix} \\ \varepsilon_{ij} &= \frac{1}{2}\left(\frac{\partial u_i}{\partial x_j} + \frac{\partial u_j}{\partial x_i}\right) \quad i = x, y, z; \quad j = x, y, z \\ \Rightarrow \varepsilon_{xy} &= -\frac{1}{2} z\alpha \\ \varepsilon_{xz} &= \frac{1}{2} y\alpha \\ &\text{rest strains zero} \end{aligned} \tag{5.2}$$

3. Stress Field

$$\begin{aligned} \sigma_{ij} &= \lambda \delta_{ij} \varepsilon_{ii} + 2G\varepsilon_{ij} \\ \Rightarrow \sigma_{xy} &= -Gz\alpha \\ \sigma_{xz} &= Gy\alpha \\ &\text{rest stresses zero} \end{aligned} \tag{5.3}$$

4. Strain Energy

$$\begin{aligned} U &= \iiint_V U_0 dv = \iiint_v \left[\int_0^{\varepsilon_{ij}} \sigma_{ij} d\varepsilon_{ij} \right] dv \\ &= \frac{1}{2} \iiint_v \sigma_{ij} \varepsilon_{ij} dv \Rightarrow \\ U &= \int_0^l \int_A \frac{1}{2} G\alpha^2 (z^2 + y^2) dA dx \\ &= \int_0^l \frac{1}{2} GI_p \alpha^2 dx \\ &\text{where} \\ I_p &= \int_A \frac{1}{2} (z^2 + y^2) dA = I_{yy} + I_{zz} = \pi \frac{r^4}{2} \end{aligned} \tag{5.4}$$

I_p is Polar moment of area of the cross-section which is the sum of two second moments of area of the cross-section about yy and zz axes.

5. Total Potential Energy

$$\pi = \int_0^l \frac{1}{2} GI_p \alpha^2 dx - ML\alpha \tag{5.5}$$

6. Take the variation of the total potential energy above

$$\delta\pi = \delta\left(\int_0^l \frac{1}{2} GI_p\alpha^2 dx\right) - ML\delta\alpha = 0$$
$$= \int_0^l GI_p\alpha\delta\alpha dx - ML\delta\alpha = 0 \tag{5.6}$$
$$= \left(\int_0^l GI_p\alpha dx - ML\right)\delta\alpha = 0$$

The above leads to

$$\int_0^l GI_p\alpha dx - ML = 0$$
$$\therefore GI_p\alpha = M \tag{5.6a}$$

Writing total twist angle as Θ which is rate of twist α times the length l, i.e., $\alpha = \frac{\Theta}{l}$, we have the torsion relation given by

$$\frac{GI_p}{l} = \frac{M}{\Theta} \tag{5.7}$$

7. Strains
From the twist rate definition $\theta_{,x} = \alpha = \frac{M}{GI_p}$, hence the strains are

$$\varepsilon_{xy} = -\frac{1}{2}z\theta_{,x} = -\frac{1}{2}\frac{Mz}{GI_p}$$
$$\varepsilon_{xz} = \frac{1}{2}y\theta_{,x} = \frac{1}{2}\frac{My}{GI_p} \tag{5.8}$$

The maximum strains occur in the outermost fiber when y or z are equal to the radius of the rod r.

$$\varepsilon_{xy\max} = -\frac{1}{2}\frac{Mr}{GI_p}$$
$$\varepsilon_{xz\max} = \frac{1}{2}\frac{Mr}{GI_p} \tag{5.9}$$

8. Stresses are obtained using (2.44) and (2.46) $\sigma_{ij} = \lambda\delta_{ij}\varepsilon_{ii} + 2G\varepsilon_{ij}$

$$\begin{aligned}\sigma_{xy} &= -Gz\theta_{,x} = -\frac{M}{I_p}z \\ \sigma_{xz} &= Gy\theta_{,x} = \frac{M}{I_p}y\end{aligned} \tag{5.11}$$

The maximum stresses likewise occur in the outermost fiber when y or z are equal to the radius of the rod r.

$$\begin{aligned}\sigma_{xy\max} &= -\frac{Mr}{I_p} \\ \sigma_{xz\max} &= \frac{Mr}{I_p}\end{aligned} \tag{5.12}$$

5.1 Finite Element Method for Torsion

We can follow the steps closely as in Chap. 4 where we dealt with the finite element method for an axially loaded bar.

Step 1: Choose the element

Figure 5.2 shows the element 1, followed by other elements that make up the structure. This element is defined in its own local coordinates denoted by $\widehat{X}\widehat{Y}\widehat{Z}$ and subjected to nodal forces $\widehat{f}_{1x}$ and $\widehat{f}_{2x}$ at the nodes 1 and 2 respectively. The displacements of the two nodes are assumed to be known during the formulation.

Step 2: Define Shape Function Matrix

There are only two nodes and their displacement values are $\widehat{\theta}_1$ and $\widehat{\theta}_2$. In order to satisfy these conditions we assume a linear function for the displacement $\widehat{\theta}(\widehat{x})$.

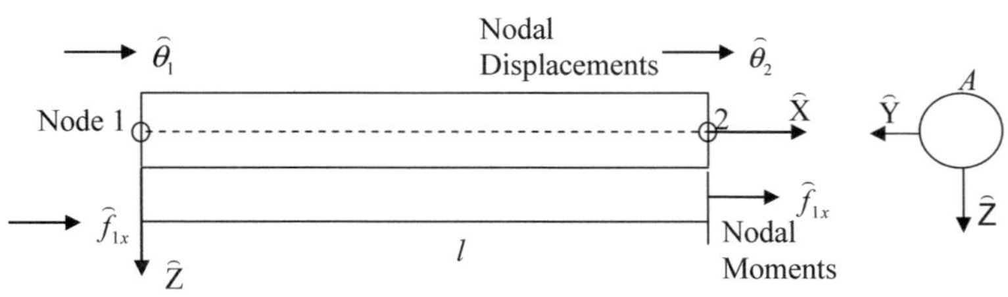

Fig. 5.2 Element 1 in local coordinates $\widehat{X}\widehat{Y}\widehat{Z}$

$$\widehat{\theta}\left(\widehat{x}\right) = a_1 + a_2\widehat{x} \tag{5.13}$$

The two constants a_1 and a_2 are determined from the conditions $\widehat{\theta}\left(\widehat{x}\right) = \widehat{\theta}_1$ at $\widehat{x} = 0$ and $\widehat{\theta}\left(\widehat{x}\right) = \widehat{\theta}_2$ at $\widehat{x} = l$. This gives

$$\begin{aligned}
\widehat{\theta}\left(\widehat{x} = 0\right) &= \widehat{\theta}_1 = a_1 \\
\widehat{\theta}\left(\widehat{x} = l\right) &= \widehat{\theta}_2 = a_1 + a_2 l \\
&= \widehat{\theta}_1 + a_2 l \Rightarrow a_2 = \frac{\widehat{\theta}_2 - \widehat{\theta}_1}{l}
\end{aligned} \tag{5.14}$$

Substituting in (5.13)

$$\begin{aligned}
\widehat{\theta}\left(\widehat{x}\right) &= \widehat{\theta}_1 + \frac{\widehat{\theta}_2 - \widehat{\theta}_1}{l}\widehat{x} \\
&= \left(1 - \frac{\widehat{x}}{l}\right)\widehat{\theta}_1 + \frac{\widehat{x}}{l}\widehat{\theta}_2
\end{aligned} \tag{5.15}$$

The above shape function is written in matrix form given by

$$\begin{aligned}
\widehat{\theta}\left(\widehat{x}\right) &= \left[1 - \frac{\widehat{x}}{l} \quad \frac{\widehat{x}}{l}\right]\left\{\begin{matrix} \widehat{\theta}_1 \\ \widehat{\theta}_2 \end{matrix}\right\} \\
&= [N_1 \quad N_2]\left\{\begin{matrix} \widehat{\theta}_1 \\ \widehat{\theta}_2 \end{matrix}\right\}
\end{aligned} \tag{5.16}$$

The shape functions in the above are same as in (4.24)

$$\begin{aligned}
N_1 &= 1 - \frac{\widehat{x}}{l} \\
N_2 &= \frac{\widehat{x}}{l}
\end{aligned} \tag{4.24}$$

We also denote the nodal degrees of freedom by

$$\widehat{d} = \begin{Bmatrix} \widehat{\theta}_1 \\ \widehat{\theta}_2 \end{Bmatrix} \tag{5.17}$$

We can then write

$$\left\{\widehat{\theta}\right\} = [N]\left\{\widehat{d}\right\} \tag{5.18}$$

Step 3: Strain-Displacement Relations

We can follow the formulation of torsion finite element in the same manner as the one dimensional axially loaded bar finite element. The engineering or nominal shear strains are used. In this case strain displacement relations are simple given by Eq. (4.2)

$$\begin{aligned} \gamma_{xy} &= -z\theta_{,x} = -z\left[N_{1,\widehat{x}} \quad N_{2,\widehat{x}}\right]\left\{\widehat{d}\right\} \\ &= -z[B]\left\{\widehat{d}\right\} \\ \gamma_{xz} &= y\theta_{,x} = y\left[N_{1,\widehat{x}} \quad N_{2,\widehat{x}}\right]\left\{\widehat{d}\right\} \\ &= y[B]\left\{\widehat{d}\right\} \end{aligned} \tag{5.19}$$

$$\begin{aligned} N_{1,\hat{x}} &= -\frac{1}{l} \\ N_{1,\hat{x}} &= \frac{1}{l} \\ \therefore [B] &= \left[-\tfrac{1}{l} \quad \tfrac{1}{l}\right] \end{aligned} \tag{5.19a}$$

Step 4: Stresses

$$\begin{aligned} \sigma_{xy} &= -Gz\theta_{,x} = -Gz[B]\left\{\widehat{d}\right\} \\ \sigma_{xz} &= Gy\theta_{,x} = -Gy[B]\left\{\widehat{d}\right\} \end{aligned} \tag{5.20}$$

$$\begin{aligned} \sigma_{xy} &= -Gz\theta_{,x} = -Gz\left[-\tfrac{1}{l} \quad \tfrac{1}{l}\right]\begin{Bmatrix} \widehat{\theta}_1 \\ \widehat{\theta}_2 \end{Bmatrix} \\ \sigma_{xz} &= Gy\theta_{,x} = -Gy\left[-\tfrac{1}{l} \quad \tfrac{1}{l}\right]\begin{Bmatrix} \widehat{\theta}_1 \\ \widehat{\theta}_2 \end{Bmatrix} \end{aligned} \tag{5.20a}$$

In view (5.13), the strains and stresses both vary linearly along the length of the element.

Step 5: Strain Energy

The strain energy in the element is

$$\begin{aligned} U &= \frac{1}{2}\left\{\hat{d}\right\}^T \int_0^l \int_A G\{(z^2+y^2)[B]^T[B]\}\left\{\hat{d}\right\} dA dx \\ &= \frac{1}{2}\left\{\hat{d}\right\}^T \int_0^l GI_p[B]^T[B]\left\{\hat{d}\right\} dx \end{aligned} \tag{5.21}$$

Step 6: Elemental Stiffness Matrix

Now we apply Eq. (4.19)

$$\frac{\partial U}{\partial\left\{\hat{d}\right\}} = \frac{\partial}{\partial\left\{\hat{d}\right\}}\left[GI_p \frac{1}{2}\left\{\hat{d}\right\}^T \int_0^l [B]^T[B]\left\{\hat{d}\right\} d\hat{x}\right] = \{M\}$$

i.e.,

$$\begin{aligned} &\left[GI_p \int_0^l [B]^T[B]\left\{\hat{d}\right\} d\hat{x}\right] = \{M\} \\ &\text{i.e.,} \\ &\left[\hat{K}\right]\left\{\hat{d}\right\} = \{M\} \end{aligned} \tag{5.22}$$

The stiffness matrix in the above is

$$[K] = GI_p \int_0^l [B]^T[B] d\hat{x} \tag{5.23}$$

Now substituting for [*B*] from (4.28)

$$[K] = GI_p \int_0^l \begin{bmatrix} N_{1,\widehat{x}} \\ N_{2,\widehat{x}} \end{bmatrix} \begin{bmatrix} N_{1,\widehat{x}} & N_{2,\widehat{x}} \end{bmatrix} d\widehat{x}$$
$$= GI_p \int_0^l \begin{bmatrix} -\frac{1}{l} \\ \frac{1}{l} \end{bmatrix} \begin{bmatrix} -\frac{1}{l} & \frac{1}{l} \end{bmatrix} d\widehat{x} \tag{5.24}$$

Performing the integration

$$[K] = \frac{GI_p}{l} \begin{bmatrix} 1 & -1 \\ -1 & 1 \end{bmatrix} \tag{5.25}$$

We notice this relation is same as axially loaded element stiffness given in (4.35), except that axial rigidity *EA* is replaced by *torsional rigidity* GI_p.

Consider one element torsion rod as in Fig. 5.1 with left end fixed and right end acted on by a moment *M*.

$$\frac{GI_p}{l} \begin{bmatrix} 1 & -1 \\ -1 & 1 \end{bmatrix} \begin{Bmatrix} 0 \\ \theta_2 \end{Bmatrix} = \begin{Bmatrix} 0 \\ M \end{Bmatrix}$$
$$\theta_2 = -\frac{Ml}{GI_p} \tag{5.7a}$$

This result is the same as Eq. (5.7) with $\theta_2 = \Theta$.

The strains and stresses vary linearly along the length from zero at the fixed end to the values at the free end. They also vary linearly with zero at the center to the maximum value at the outer radius at a given distance along the axis, see Eq. (5.20a).

We notice that the axially loaded bar and rod in torsion both are similar in their formulation, the axial load *P* in the bar is replaced by twisting moment *M*, the axial deformation *u* is replaced by the twist θ and the axial rigidity of the bar *EA* by torsional rigidity GI_p.

5.2 Two Elements and Stiffness Matrix Assembly

We have now three nodes and two elements, element 1 with nodes 1 and 2 with area A_1, length l_1 and element 2 with nodes 2 and 3 with area A_2, length l_2, see Fig. 5.3. The Moment *M* is applied at node 3. For element 1,

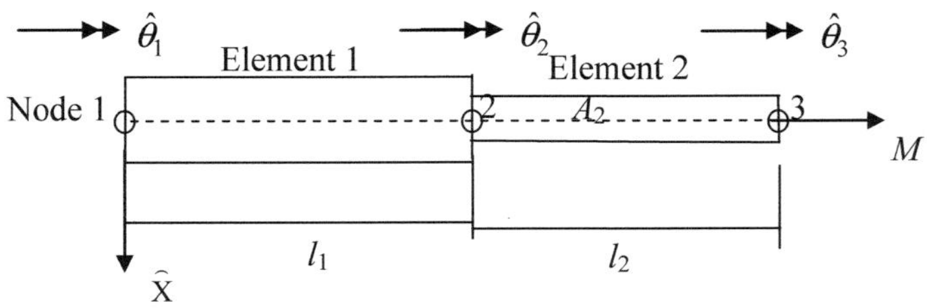

Fig. 5.3 Two element model

$$\left[\hat{K}_1\right]\left\{\hat{d}_1\right\} = \{0\}$$
$$\frac{GI_{p1}}{l_1}\begin{bmatrix} 1 & -1 \\ & 1 \end{bmatrix}\begin{Bmatrix} \hat{\theta}_1 \\ \hat{\theta}_2 \end{Bmatrix} = \begin{Bmatrix} 0 \\ 0 \end{Bmatrix} \tag{5.26}$$

For element 2,

$$\left[\hat{K}_2\right]\left\{\hat{d}_2\right\} = \frac{GI_{p2}}{l_2}\begin{bmatrix} 1 & -1 \\ & 1 \end{bmatrix}\begin{Bmatrix} \hat{\theta}_2 \\ \hat{\theta}_3 \end{Bmatrix} = \begin{Bmatrix} 0 \\ M \end{Bmatrix} \tag{5.27}$$

For the rod we now assemble the elemental stiffness matrices to get the global stiffness matrix.

$$\begin{bmatrix} \frac{GI_{p1}}{l_1} & -\frac{GI_{p1}}{l_1} & \\ & \frac{GI_{p1}}{l_1} + \frac{GI_{p2}}{l_2} & -\frac{GI_{p2}}{l_2} \\ & & \frac{GI_{p2}}{l_2} \end{bmatrix}\begin{Bmatrix} \theta_1 = 0 \\ \theta_2 \\ \theta_3 \end{Bmatrix} = \begin{Bmatrix} 0 \\ 0 \\ M \end{Bmatrix} \tag{5.28}$$

Taking away the column and row of node 1

$$\begin{bmatrix} \frac{GI_{p1}}{l_1} + \frac{GI_{p2}}{l_2} & -\frac{GI_{p2}}{l_2} \\ & \frac{GI_{p2}}{l_2} \end{bmatrix}\begin{Bmatrix} \theta_2 \\ \theta_3 \end{Bmatrix} = \begin{Bmatrix} 0 \\ M \end{Bmatrix} \tag{5.29}$$

$$\begin{aligned} &\left(\frac{GI_{p1}}{l_1} + \frac{GI_{p2}}{l_2}\right)\theta_2 - \frac{GI_{p2}}{l_2}\theta_3 = 0 \\ &-\frac{GI_{p2}}{l_2}\theta_2 + \frac{GI_{p2}}{l_2}\theta_3 = M \end{aligned} \tag{5.29a}$$

First determine θ_2 in terms of θ_3

$$\theta_2 = \frac{\frac{GI_{p2}}{l_2}}{\frac{GI_{p1}}{l_1} + \frac{GI_{p2}}{l_2}} \theta_3$$

Eliminate θ_2

$$\left(-\frac{GI_{p2}}{l_2} \frac{\frac{GI_{p2}}{l_2}}{\frac{GI_{p1}}{l_1} + \frac{GI_{p2}}{l_2}} + \frac{GI_{p2}}{l_2} \right) \theta_3 = M$$

The twist at the right end is

$$\theta_3 = M \frac{\frac{GI_{p1}}{l_1} + \frac{GI_{p2}}{l_2}}{\frac{GI_{p1}}{l_1} \frac{GI_{p2}}{l_2}} \tag{5.30}$$

If both the elements have area I_p and length 1/2 l the above gives $\theta_3 = \frac{Ml}{GI_p}$ as before.

Worked Example 5.1

Consider a rod of length 1 m and diameter 0.1 m acted subjected to a 10,000 Nm as shown in Fig. Worked Example 5.1.

Analytical Solution: From Eq. (5.7) or one element solution using FEM

$$\Theta = \frac{Ml}{GI_p} = \frac{10{,}000 \times 1}{79.3 \times 10^9 \frac{\pi \times 0.05^4}{2}} = 0.01284 \text{ rad} = 0.7357\circ$$

Maximum shear Stress τ in outer fiber from Eq. (5.12)

$$\tau = \frac{Mr}{I_p} = \frac{10{,}000 \times 0.05}{\frac{\pi \times 0.05^4}{2}} = \frac{500}{9.8175 \times 10^{-6}} = 50.929 \text{ MPa}$$

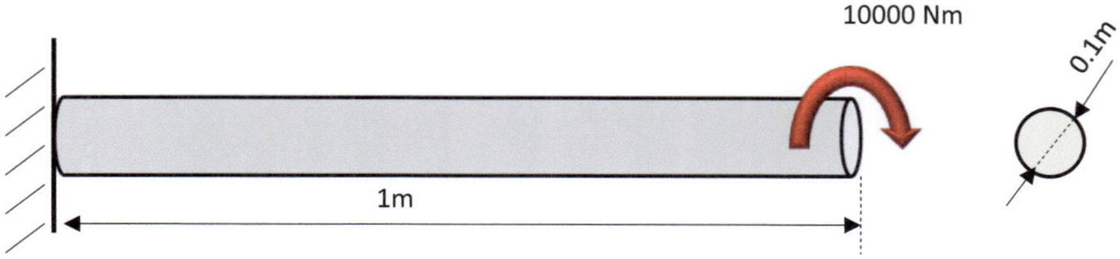

Fig. Worked Example 5.1 One dimensional rod

Worked Example 5.2

Determine the diameter d of a shaft transmitting 50 HP at 200 rpm by not allowing the shear stress to cross 50 MPa. The shaft length is 2.5 m.

Using $HP = \frac{2\pi NT}{45{,}000}$

$$50 = \frac{2\pi \times 200T}{45{,}000} \Rightarrow T = \frac{45{,}000 \times 50}{400\pi} = 1790.5 \text{ Nm}$$

$$\therefore 50 \times 10^6 = \frac{1790.5d}{2\frac{\pi \times d^4}{64}} = \frac{1790.5 \times 10.186}{d^3} = \frac{18237.9}{d^3}$$

$$\Rightarrow d^3 = \frac{18237.9}{50 \times 10^6} = 0.00036476$$

$$d = 0.0714 \text{ m} = 7.14\,\text{cm}$$

Worked Example 5.3

A propeller shaft 30 cm diameter transmits a torque 200,000 Nm at 500 rpm. Over a length of 2.5 m, what is the twist angle it suffers? Determine also the maximum shear stress the propeller shaft is subjected to. The shear modulus G = 80,000 MPa.

$$\Theta = \frac{Ml}{GI_p} = \frac{200{,}000 \times 2.5}{80 \times 10^9 \frac{\pi \times 0.15^4}{2}} = \frac{500{,}000}{80 \times 10^9 \times 0.0007952} = 0.0079 \text{ rad}$$

$$\tau = \frac{Mr}{I_p} = \frac{200{,}000 \times 0.15}{\frac{\pi \times 0.15^4}{2}} = \frac{30{,}000}{7.952 \times 10^{-4}} = 37.726 \text{ MPa}$$

Worked Example 5.4

A propeller shaft 5.08 cm diameter transmits a torque 2685.971 Nm. Over a length of 0.6096 m, what is the twist angle it suffers? Determine also the maximum shear stress the propeller shaft is subjected to. The shear modulus G = 82,031.25 MPa.

$$\Theta = \frac{Ml}{GI_p} = \frac{2685.971 \times 0.6096}{82{,}031 \times 10^6 \frac{\pi \times 0.0254^4}{2}}$$

$$= \frac{1637.368}{82{,}031 \times 65.381} = 0.0003053\,\text{rad} = 0.01749\,\circ$$

$$\tau = \frac{Mr}{I_p} = \frac{2685.971 \times 0.0254}{\frac{\pi \times 0.0254^4}{2}} = 104.31\,\text{MPa}$$

Worked Example 5.5

A cantilever rod 1 m long and 10 cm dia is subjected to an end torque 10,000 Nm. Determine by FEM the twist angle and maximum shear stress. Use $G = 79.3 \times 10^9$ Pa.

The analytical solutions for the total twist and maximum shear stress are

$$\Theta = \frac{Ml}{GI_p} = \frac{10{,}000 \times 1}{79.3 \times 10^9 \frac{\pi \times 0.05^4}{2}} = 0.01284 \text{ rad} = 0.73568\,^\circ$$

$$\tau = \frac{Mr}{I_p} = \frac{10{,}000 \times 0.05}{\frac{\pi \times 0.05^4}{2}} = \frac{500}{9.8175 \times 10^{-6}} = 50.929 \text{ MPa}$$

While one dimensional axially loaded elements are available in many codes (so also for bending), unfortunately we do not have such elements for torsion. Therefore, a large number of solid elements are required to make the FE model. The number of elements used in such a model shown in Fig. Worked Example 5.1.1 are 96,600 containing 101,926 nodes.

The twist angle is also not directly available and it can be determined by identifying from linear displacements as shown in Fig. Worked Example 5.5.2. Maximum Linear Displacement is found to be 0.63 mm. The angle of twist is therefore

$r\Theta = 0.63 \Rightarrow \Theta = \frac{0.63}{50} = 0.0126$ rad $= 0.722\,^\circ$. This is in good agreement with the closed form solution.

This angle can also be determined by superimposing the original mesh on the deformed mesh and determining the angle of any one radial line, see Fig. Worked Example 5.5.3.

The maximum stresses are given by Eq. (5.12)

$$\sigma_{xy\,\max} = -\frac{Mr}{I_p}\ \sigma_{xz\,\max} = \frac{Mr}{I_p}$$

These stresses are shown in Fig. Worked Example 5.5.4 and Fig. Worked Example 5.5.5. They are in good agreement with the theoretical values.

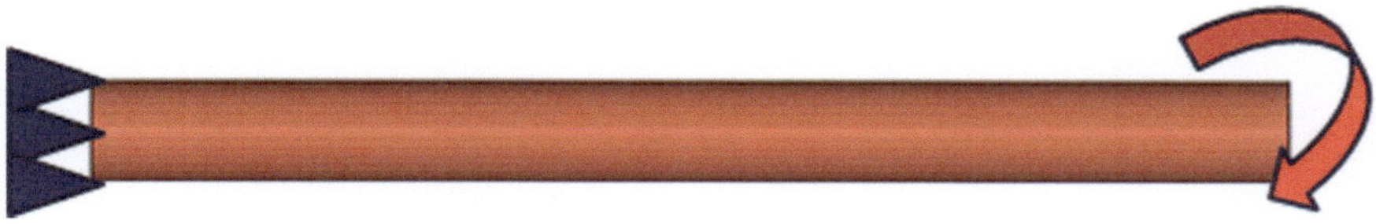

Fig. Worked Example 5.5.1 A Cantilever Rod

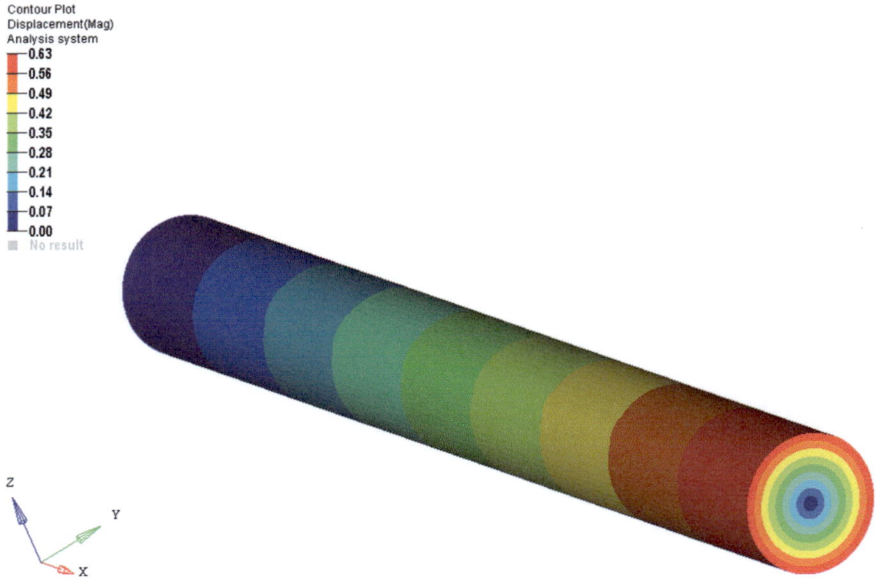

Fig. Worked Example 5.5.2 Linear displacement plot

Fig. Worked Example 5.5.3 Twist angle

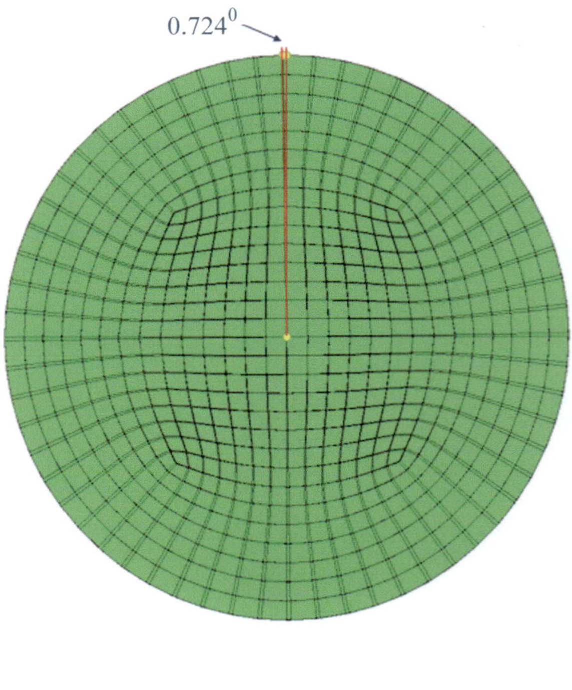

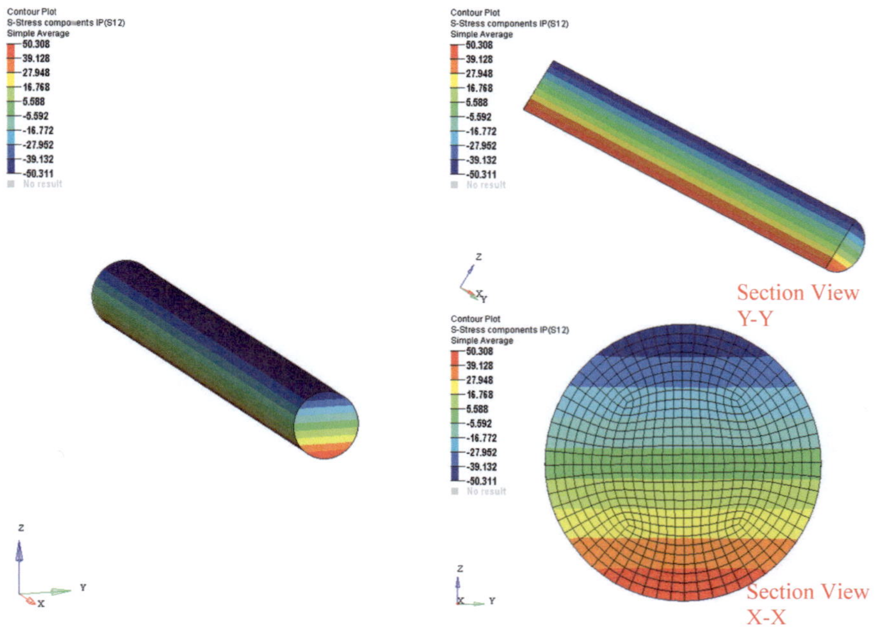

Fig. Worked Example 5.5.4 Shear stress $\sigma_{xy} = 50.3\,\text{MPa}$

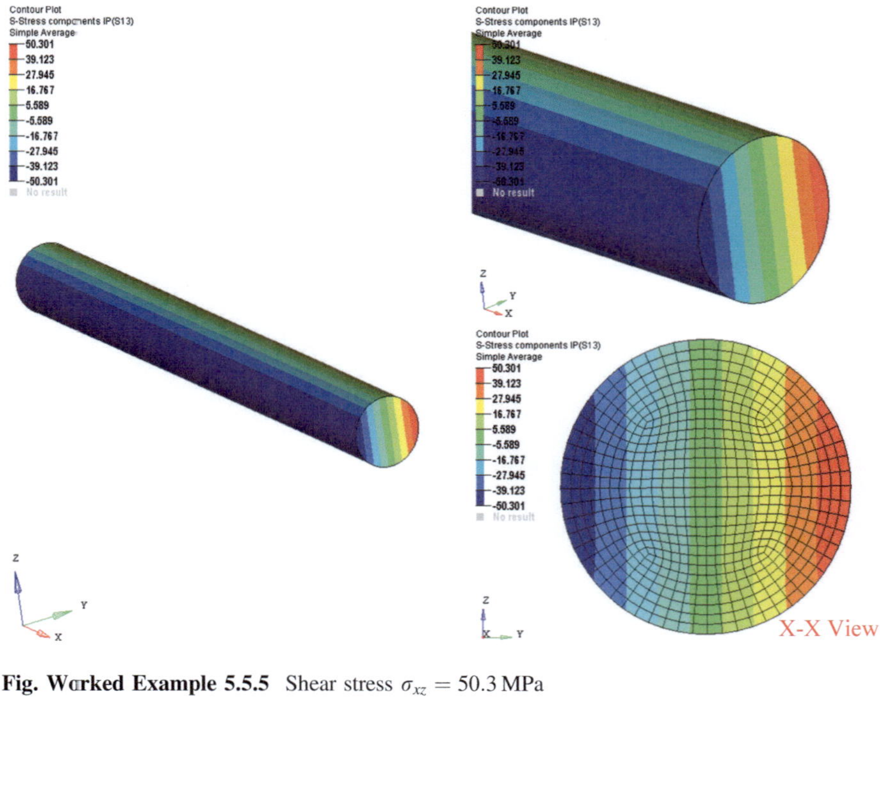

Fig. Worked Example 5.5.5 Shear stress $\sigma_{xz} = 50.3\,\text{MPa}$

5.3 Ritz Method for Torsion

We assume the deflection or shape function to be

$$\begin{aligned}\theta(x) &= ax\\ \frac{d\theta}{dx} &= \theta_{,x} = a\end{aligned} \tag{5.31}$$

The total potential energy π is

$$\begin{aligned}\pi = U - W &= \int_0^l \left(\frac{1}{2}GI_p a^2\right) dx - Mal\\ &= \frac{1}{2}GI_p l a^2 - Mal\end{aligned} \tag{5.32}$$

Minimize the total potential energy by making its derivative with respect to a to be zero

$$\frac{\partial \pi}{\partial a} = aGI_p l - Ml = 0 \Rightarrow a = \frac{M}{GI_p} \tag{5.33}$$

The deflection $\theta(x)$ is then given by

$$\theta(x) = \frac{M}{GI_p}x$$

The tip elongation is then

$$\theta_{\text{tip}} = \frac{Ml}{GI_p}$$

5. Exercises

5.1 Describe the displacement field used for torsion of circular shafts.
5.2 For a circular shaft subjected to a torque where is the maximum shear stress occurring along the length of the shaft and in the radius.
5.3 Derive the strain and stress along the length of a shaft subjected to a twisting moment and show that they remain the same along the axis but vary from radius to outermost fiber in a linear manner.
5.4 Derive the finite element method for torsion of circular rods. Make sure you have a constant twist slope along the axis.

5.5 Determine the diameter of a solid steel shaft that is required to transmit 450 HP at 300 rpm if the twist is not to exceed one degree over a length of 2.5 m.

5.6 In the above example what will be the diameter if an additional condition that the stress should not exceed 40 MPa.

5.7 An engine 80 HP transmits power at 100 rpm through a steel propeller shaft. Determine the size for 30 % extra power over the mean if the maximum shear stress is not to exceed 80 MPa. Then what would be the twist over a length of 3 m.

5.8 A ship having a single propeller shaft 30 cm diameter and running at 160 rpm is re-engineered with turbines driving two equal dia propeller shafts at 800 rpm and developing 50 % more horse power. Allow 10 % more shear stress and determine the new design propeller shafts diameter.

5.9 A steel shaft 2 m long is tapered from 7.5 to 10 cm. The shaft is fixed at the larger diameter end and a torque of 3500 Nm is applied at the other end. Use two element uniform dia model and find the total angle of twist and maximum shear stress.

5.10 Use Ritz method and obtain the angular displacement and stress in the above problem.

Chapter 6
Bending of a Beam

All structures are three dimensional, we make convenient approximations to get a quick understanding and simplify methods of determining stress and strain at a point under given loading. The simplest of them are one-dimensional structures, where only one space coordinate is accounted while defining the structural behavior. This is permissible when the cross-sectional dimensions of a structure are small compared to the length dimension.

We also make assumptions on the deformation field of the structure, e.g., if we stretch such a one-dimensional structure, we have a bar; if we twist this structure by a pure moment, then we have a rod (shaft in pure torsion) and another most commonly used simple one-dimensional structure is a beam with a load in transverse or lateral direction.

We will consider a beam in its simplest form, see Fig. 6.1, and derive a simple theory called technical beam theory from fundamentals. This will also be same as Euler-Bernoulli theory. We recognize the beam to be lying on the x-axis (only one dimension); the cross-sectional dimensions enter into the theory through the derivation. We will see these parameters entering the governing equation to be the area A and second moment of area I of the cross-section.

Consider the cross-section at a distance x, this cross-section is in a plane normal to the x axis before bending and after bending, this plane is assumed to remain plane (no warping or distortions) and rotate in such a way that it still remains normal to the bent (elastic) axis. This is what Leonardo Da Vinci has observed long ago and firmed up by Euler and Bernoulli.

By approximating the beam and forcing the cross-section at distance x, we are writing an approximate solution rather than solving the 15 equations of theory of elasticity derived before. This approximation is so good it gives us almost correct solution for slender beams when compared with experimental results. Through this approach we have an approximate but working theory. From Fig. 6.1 the displacement field of a point at coordinates x, y, z of the beam can be written as

J.S. Rao, *Simulation Based Engineering in Solid Mechanics*,
DOI 10.1007/978-3-319-47614-8_6

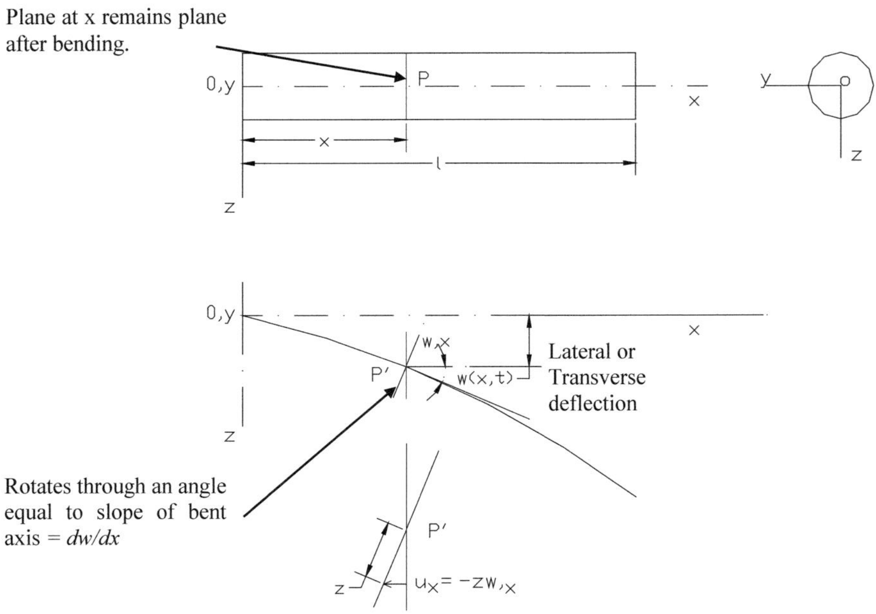

Fig. 6.1 Technical theory of Euler beam

$$\begin{aligned} u_x &= -z\, w_{,x} \\ u_y &= 0 \\ u_z &= w(x,t) \end{aligned} \tag{6.1}$$

Notice that a comma followed by suffix x as before denotes derivative with respect to x ($_{,x} = d/dx$). The beam axis has no strain at all and therefore the elastic axis is also a neutral axis.

Once the displacement field is assumed we have violated the exact theory of elasticity; however this allows us to obtain a simple solution because we can obtain the strains directly from strain displacement relations.

$$\begin{aligned} \varepsilon_{xx} &= \frac{\partial u}{\partial x} = -z\frac{\partial^2 w}{\partial x^2} & \varepsilon_{yz} &= \frac{1}{2}\left(\frac{\partial v}{\partial z} + \frac{\partial w}{\partial y}\right) = \varepsilon_{zy} = 0 \\ \varepsilon_{yy} &= \frac{\partial v}{\partial y} = 0 & \varepsilon_{zx} &= \frac{1}{2}\left(\frac{\partial w}{\partial x} + \frac{\partial u}{\partial z}\right) = \varepsilon_{xz} = 0 \\ \varepsilon_{zz} &= \frac{\partial w}{\partial z} = 0 & \varepsilon_{xy} &= \frac{1}{2}\left(\frac{\partial u}{\partial y} + \frac{\partial v}{\partial x}\right) = \varepsilon_{yx} = 0 \end{aligned} \tag{6.2}$$

Only one normal strain along the x axis exists; rest strains are zero. We have not taken into account cognizance of Poisson's ratio.

The stress field is given by

$$\sigma_{ij} = \lambda\delta_{ij}\varepsilon_{ii} + 2G\varepsilon_{ij}$$
$$\rightarrow \sigma_{xx} = E\varepsilon_{xx} = -Ez\frac{\partial^2 w}{\partial x^2} \quad (6.3)$$

Likewise there is only one normal stress and no other stresses including any shear stress. An approximate theory built from simple assumption for long slender beam and yet it is sufficiently accurate that this theory is still used from 1750.

We will first use the above Eq. (6.3) to get the stresses and understand the equilibrium conditions. Stress at distance z in plane P at distance x is

$$\sigma_{xx} = -Ez\frac{\partial^2 w}{\partial x^2} \quad (6.3a)$$

When z is positive below yy axis in Fig. 6.2, the stress is negative that is compressive and in the fibers above yy axis z is negative and therefore stress is positive or tensile in nature. The stress on yy axis where $z = 0$, the stress is zero, therefore it is a neutral axis. Let the beam be of width b, thickness t. Force on element at distance z thickness dz and width b (area $b\ dz$) is

$$dF = -Eb\frac{d^2 w}{dx^2}zdz \quad (6.4)$$

Taking moment of this elemental force and integrating over the entire thickness from $-1/2\ t$ to $+1/2\ t$, we get the familiar bending moment of the beam.

$$\begin{aligned} M_{yy} &= \int_{-\frac{1}{2}t}^{+\frac{1}{2}t} -Eb\frac{d^2 w}{dx^2}z^2 dz \\ &= \left[-\frac{1}{3}Eb\frac{d^2 w}{dx^2}z^3\right]_{-\frac{1}{2}t}^{+\frac{1}{2}t} \\ &= -\frac{1}{12}Ebt^3\frac{d^2 w}{dx^2} = -EI\frac{d^2 w}{dx^2} \end{aligned} \quad (6.5)$$

Since there is no other internal force, this beam is said to be under pure bending with a bending moment about yy axis. Externally, the transverse load applied on the beam causes a moment at a section distance x and that moment will counter the internal bending moment caused by the stress field. Alternatively, the traction loads

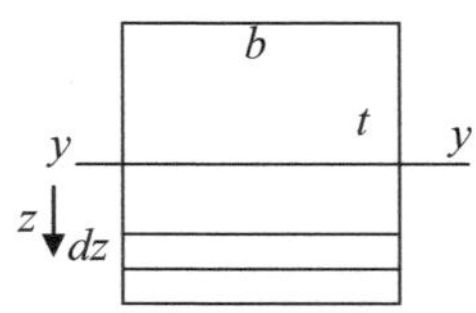

Fig. 6.2 Elemental force

applied on the beam are resisted by the stress field given in Eq. (6.5). *EI* is the flexural rigidity of the beam.

Euler's greatness lies in imagining a beam as he defined from real life three dimensional structures around. A beam with vertical loads, many even today conceive that the predominant stress is in the direction of the load, actually bends because of bending moment that arises from the loading, not so much because of the vertical load. The stress is axial even though the load is vertical.

While this theory holds good even today, remember it is an approximate theory, where we have assumed the pattern (or modeled) in which bending takes place and derived the rest of the theory. There is an effect of shear in the vertical plane which is very small compared with bending moment and Timoshenko derived an equation taking into account of shear in 1926.

There have been several attempts to make further refinements subsequently, e.g., Volterra to satisfy all the requirements stipulated by Theory of Elasticity consistent with one dimensional structure. When we consider a one D structure it is found sufficient to include Timoshenko shear effect and ignore the rest.

Worked Example 6.1

Derive the tip deflection for a cantilever with an end moment *M*.

The bending moment relation from Eq. (6.5) is given by

$$EI\frac{d^2y}{dx^2} = -M$$

Integrating twice

$$EI\frac{dy}{dx} = -Mx + A$$

$$EIy = -\frac{1}{2}Mx^2 + Ax + B$$

At $x = 0$, both the deflection and slope are zero, therefore

$$EIy = -\frac{1}{2}Mx^2$$

$$y = -\frac{1}{2}\frac{M}{EI}x^2$$

The deflection is maximum at the tip $x = l$.

$$y_{\max} = \frac{1}{2}\frac{Ml^2}{EI}$$

Worked Example 6.2

Derive the tip deflection for a cantilever with an end force *P*.

The bending moment at a distance from free end is $-Px$ and the relation from Eq. (6.5) is given by

$$EI\frac{d^2y}{dx^2} = -Px$$

Integrating twice

$$EI\frac{dy}{dx} = -\frac{1}{2}Px^2 + A$$
$$EIy = -\frac{1}{6}Px^3 + Ax + B$$

At $x = l$, both the deflection and slope are zero, therefore

$$0 = -\frac{1}{2}Pl^2 + A \Rightarrow A = \frac{1}{2}Pl^2$$
$$0 = -\frac{1}{6}Pl^3 + Al + B \Rightarrow B = -Pl^3\left(\frac{1}{2} - \frac{1}{6}\right) = -\frac{1}{3}Pl^3$$
$$\therefore y = -\frac{1}{6}Px^3 + \frac{1}{2EI}Pl^2x - \frac{1}{3EI}Pl^3$$

The deflection is maximum at the tip $x = 0$.

$$y|_{x=l} = -\frac{1}{3EI}Pl^3$$

6.1 Bending by Energy Method

We will introduce the energy method which is most fundamental to all elasticity problems and will be recalled when we look at Finite Element Methods. The strain energy of the whole beam U can be written as

$$U = \frac{1}{2}\iiint_v \sigma_{ij}\varepsilon_{ij}dv = \frac{1}{2}b\int_0^L\int_{-\frac{1}{2}t}^{\frac{1}{2}t} \sigma_{xx}\varepsilon_{xx}dzdx \tag{6.6}$$

Work W of loading $q(x)$ assumed to act on the centroidal axis is

$$W = \frac{1}{2}\int_0^L q(x)w(x)dx \tag{6.7}$$

Total potential energy π is

$$\begin{aligned}\pi = U - W &= \frac{1}{2} b \int_0^L \int_{-\frac{1}{2}t}^{\frac{1}{2}t} \sigma_{xx}\varepsilon_{xx} dz dx - \frac{1}{2} \int_0^L q(x)w(x)dx \\ &= \frac{1}{2} b \int_0^L \int_{-\frac{1}{2}t}^{\frac{1}{2}t} \left(Ez^2\right)\left(\frac{d^2w}{dx^2}\right)^2 dz dx - \frac{1}{2}\int_0^L q(x)w(x)dx \\ &= \int_0^L \left[\frac{1}{2} EI \left(\frac{d^2w}{dx^2}\right)^2 - qw\right] dx = \int_0^L F dx\end{aligned} \tag{6.8}$$

where F is a functional given by

$$F = \frac{1}{2} EI \left(\frac{d^2w}{dx^2}\right)^2 - qw \tag{6.9}$$

The strain energy in the beam in (6.6) can also be directly written from work of bending moment through the elemental angle of rotation.

To derive the governing differential equation for the beam, we use, Euler-Lagrange equation given in Chap. 3

$$-\frac{d^2}{dx^2}\left(\frac{\partial F}{\partial w''}\right) + \frac{d}{dx}\left(\frac{\partial F}{\partial w'}\right) - \left(\frac{\partial F}{\partial w}\right) = 0 \tag{3.16}$$

where $w' = \frac{dw}{dx}, \ldots$ Using F in (6.9) we first form the required partial derivatives

$$\begin{aligned}\frac{\partial F}{\partial w''} &= EI\left(\frac{d^2w}{dx^2}\right) \\ \frac{\partial F}{\partial w'} &= 0 \\ \frac{\partial F}{\partial w} &= -q\end{aligned} \tag{6.10}$$

Substituting the above in (3.16) we get the governing differential equation of the beam.

$$\frac{d^2}{dx^2}\left[EI\left(\frac{d^2w}{dx^2}\right)\right] = q \tag{6.11}$$

Worked Example 6.3

A simply supported beam carries a concentrated vertical load P at its middle. Determine the deflection of the beam from integration of the bending moment relation and by energy method.

By integration: The two supports have a reaction force equal to 1/2 P. Therefore the bending moment at a position x up to the middle is 1/2 Px. Then

$$EI\frac{d^2y}{dx^2} = \frac{1}{2}Px$$

Integrating

$$\begin{aligned} EI\frac{dy}{dx} &= \frac{1}{4}Px^2 + A \\ EIy &= \frac{1}{12}Px^3 + Ax + B \end{aligned}$$

We utilize the symmetry and assume at x = 0, y = 0 and the slope is zero at x = 1/2 l; Therefore

$$\begin{aligned} EI\frac{dy}{dx} &= 0 = \frac{1}{16}Pl^2 + A \Rightarrow A = -\frac{1}{16}Pl^2 \\ EIy &= \frac{1}{12}Px^3 + Ax + B \Rightarrow B = 0 \\ \therefore y &= \frac{1}{12EI}Px^3 - \frac{1}{16}Pl^2x \end{aligned}$$

Maximum deflection at the middle is

$$y|_{x=\frac{1}{2}l} = \frac{1}{96EI}Pl^3 - \frac{1}{32}Pl^3 = -\frac{1}{48}Pl^3$$

By Energy Method:

Let us assume the shape function to be $y_1 = a\,\sin\frac{\pi x}{l}$ that satisfies the kinematic boundary conditions y = 0 and natural boundary conditions $\frac{d^2y}{dx^2} = 0$ at the supports.

The total potential energy is

$$\begin{aligned} \pi &= \frac{EI}{2}\int_0^l \left(\frac{d^2y}{dx^2}\right)^2 dx - Py|_{x=\frac{1}{2}l} \\ &= \frac{EI}{2}a^2\left(\frac{\pi}{l}\right)^4 \int_0^l \sin^2\frac{\pi x}{l}dx - Pa \\ &= \frac{EI}{2}a^2\left(\frac{\pi}{l}\right)^4 \frac{l}{2} - Pa \end{aligned}$$

Minimizing

$$\frac{\partial \pi}{\partial a} = \frac{EI}{2} 2a \left(\frac{\pi}{l}\right)^4 \frac{l}{2} - P = 0 \Rightarrow a = \frac{2Pl^3}{\pi^4 EI}$$

Then

$$y_1 = \frac{2Pl^3}{\pi^4 EI} \sin \frac{\pi x}{l}$$

Also

$$y|_{x=\frac{1}{2}l} = \frac{2Pl^3}{\pi^4 EI} = \frac{1}{48.7045} \frac{Pl^3}{EI}$$

The energy method result is close to the classical integration method result.

6.2 Beam with Axial Load (Beam-Column)

Consider a beam with axial compressive force S in Fig. 6.3.

The work done by axial compressive force S is

$$W_S = \int_0^L \frac{1}{2} S w_{,x}^2 dx \tag{6.12}$$

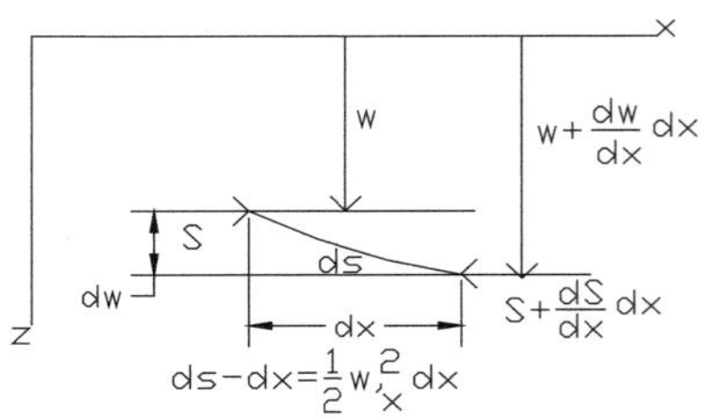

Fig. 6.3 Beam with axial compressive force

Hence

$$\pi = U - W = \frac{1}{2}b\int_{0}^{L}\int_{-\frac{1}{2}t}^{\frac{1}{2}t}\sigma_{xx}\varepsilon_{xx}dzdx - \int_{0}^{L}\frac{1}{2}Sw_{,x}^{2}dx - \frac{1}{2}\int_{0}^{L}q(x)w(x)dx$$

$$= \frac{1}{2}b\int_{0}^{L}\int_{-\frac{1}{2}t}^{\frac{1}{2}t}(Ez^{2})\left(\frac{d^{2}w}{dx^{2}}\right)^{2}dzdx - \int_{0}^{L}\frac{1}{2}Sw_{,x}^{2}dx - \frac{1}{2}\int_{0}^{L}q(x)w(x)dx \qquad (6.13a)$$

Integrating and simplifying, the above becomes

$$\pi = \int_{0}^{L}\left[\frac{1}{2}EI\left(\frac{d^{2}w}{dx^{2}}\right)^{2} - \frac{1}{2}Sw_{,x}^{2} - qw\right]dx = \int_{0}^{L}Fdx \qquad (6.13b)$$

Using

$$F = \frac{1}{2}EI\left(\frac{d^{2}w}{dx^{2}}\right)^{2} - \frac{1}{2}S\left(\frac{dw}{dx}\right)^{2} - qw \qquad (6.14)$$

Then,

$$\frac{\partial F}{\partial w''} = EI\left(\frac{d^{2}w}{dx^{2}}\right)$$

$$\frac{\partial F}{\partial w'} = -S\left(\frac{dw}{dx}\right)$$

$$\frac{\partial F}{\partial w} = -q$$

$$-\frac{d^{2}}{dx^{2}}\left(\frac{\partial F}{\partial w''}\right) + \frac{d}{dx}\left(\frac{\partial F}{\partial w'}\right) - \left(\frac{\partial F}{\partial w}\right) = 0$$

$$\begin{aligned} -\frac{d^{2}}{dx^{2}}\left[EI\left(\frac{d^{2}w}{dx^{2}}\right)\right] - \frac{d}{dx}\left(S\frac{dw}{dx}\right) &= q \\ \frac{d^{2}}{dx^{2}}\left[EI\left(\frac{d^{2}w}{dx^{2}}\right) + Sw\right] &= -q \end{aligned} \qquad (6.15)$$

Clearly from simple physical reasoning, the effects of axial load S on w will directly depend on the deflection that $q(x)$ induces.

First let us consider the case of a tension applied instead of compression. Equation (6.15) becomes

$$\frac{d^2}{dx^2}\left[EI\left(\frac{d^2w}{dx^2}\right) - Sw\right] = -q \tag{6.15a}$$

Letting $\lambda = \sqrt{\frac{S}{EI}}$, dividing the above equation by EI

$$\frac{d^2}{dx^2}\left[\left(\frac{d^2w}{dx^2} - \lambda^2 w\right)\right] = -\frac{q}{EI} \tag{6.15b}$$

The solution for homogeneous part of the above equation is

$$w = A\sin\lambda x + B\cos\lambda x + Cx + D$$

Satisfaction of the boundary conditions for the simply supported beam namely w and w'' are zero at $x = 0$ and L then

$$\begin{aligned} &B + D = 0 \\ &A\sin\lambda L + B\cos\lambda L + CL + D = 0 \\ &B = 0 \\ &A\sin\lambda L + B\cos\lambda L = 0 \end{aligned}$$

Solution of these gives the nontrivial solution $B = C = D = 0$ and $A\sin\lambda L = 0$. Therefore $\lambda = \frac{i\pi}{L}$ $i = 1, 2,$ and

$$w = \sin\frac{i\pi x}{L} \tag{6.16}$$

Substituting the above in (6.15b) for the homogeneous part, we have

$$\begin{aligned} &\left(\frac{i\pi}{L}\right)^4 \sin\frac{i\pi x}{L} + \lambda^2\left(\frac{i\pi}{L}\right)^2 \sin\frac{i\pi x}{L} = 0 \\ &\left(\frac{\pi}{L}\right)^2 - \lambda^2 = 0 \Rightarrow \lambda^2 = \left(\frac{\pi}{L}\right)^2 \Rightarrow S = EI\left(\frac{\pi}{L}\right)^2 \end{aligned} \tag{6.17}$$

S in the above equation gives the Euler Buckling load.

Worked Example 6.4

Determine Euler Buckling load for a cantilever using an equivalent length approach.

The critical load for a cantilever can be expressed in terms of the critical buckling load for the pinned column as given in Fig. Worked Example 6.4

From Euler length for buckling of a pinned beam that represents the total bent shape, L_E we can derive appropriate equivalent lengths for other boundary conditions. For a cantilever, the displacement and slope are zero at the fixed end and a finite value of slope at the free end. From L_E we identify the span as shown. The span of the equivalent cantilever is equivalent to 1/2 of Euler span, therefore

Fig. Worked Example 6.4 Equivalent length of a Cantilever column

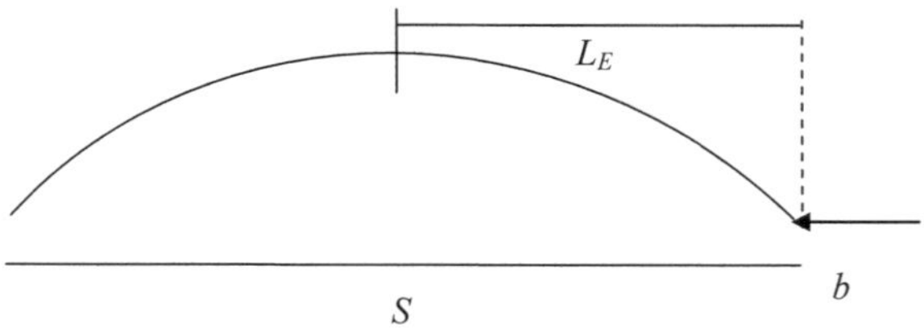

$$S_{\text{Critical for Cantilever}} = \frac{\pi^2 EI}{L_E^2} = \frac{\pi^2 EI}{4L^2} = \frac{1}{4} S_{\text{Critical for Pinned}}$$

Worked Example 6.5

A 2 m long pin ended column is of square cross section. Young's modulus of the material is $E = 12$ GPa and maximum stress allowed is 12 MPa for compression parallel to the grain. A factor of safety of 2.5 is required in computing Euler's critical load for buckling. Determine the size of the cross section if the column is to safely support a load $S = 100$ kN.

Let b be the side of the square cross-section. From (6.17)

$$250{,}000 = 12 \times 10^6 \times \frac{b^4}{12}\frac{\pi^2}{4} \Rightarrow b^4 = \frac{250{,}000 \times 48}{12 \times 10^9 \pi^2} = \frac{1}{1000\pi^2}$$

$$b = 0.1003 \text{ m or } 10 \text{ cm}$$

6.3 Strength of Materials Approach

In the previous sections we looked at fundamentals of elasticity from first principles. The fifteen coupled partial differential equations for 3 displacements, 6 each stress and strain components are too complex for any solution; therefore engineers looked initially at simpler approaches to find solutions to practical problems, called *Strength of Materials*. Let us consider a long slender structure, we called this *beam*.

Instead of getting into the detailed deformation field at a point, let us take a global element dx of the beam at distance x and consider its equilibrium. We begin with the existence of bending moment M_x and shear force V_x at distance x to keep the element in equilibrium against the applied load q. Note the convention, on the right face the shear force and the applied load are positive downwards and bending moment is positive in counter clockwise direction.

Figure 6.4 gives the equilibrium diagram of the beam element. The moment and force equilibrium equations are

$$
\begin{aligned}
-M_x + (M_x + dM_x) - V_x dx + q dx\left(\frac{dx}{2}\right) &= 0 \\
V_x - (V_x + dV_x) - q dx = 0
\end{aligned}
\tag{6.18}
$$

Neglecting second order terms and rearranging, we get

$$
\begin{aligned}
\frac{dM_x}{dx} &= V_x \\
\frac{dV_x}{dx} &= -q
\end{aligned}
\tag{6.19}
$$

We notice that the bending moment and shear force are related between themselves, while the shear force is directly connected with the loading.

Now consider the beam geometry after deformation from pure bending, see Fig. 6.5. The beam bends with its neutral axis under the couple M with a *curvature* $1/\rho$. Let the element dx making an angle $d\theta$ that is the plane section before bending rotates through an angle $d\theta$ and remains plane after bending. The fiber at distance z from neutral axis elongates by $zd\theta$, giving the strain

$$\varepsilon_{xx} = \frac{zd\theta}{dx} = \frac{zd\theta}{\rho d\theta} = \frac{z}{\rho} \tag{6.20}$$

Notice all the fibers on the concave side are in tension and on the convex side are in compression. Also

$$\frac{d\theta}{dx} = \frac{1}{\rho} \tag{6.21}$$

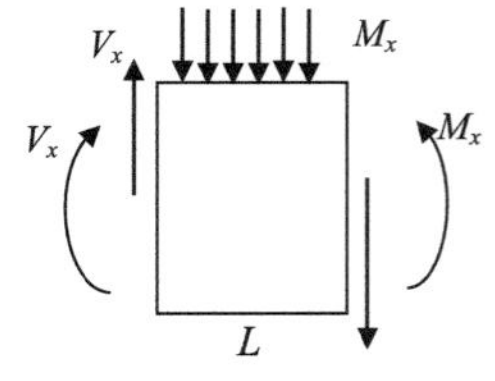

Fig. 6.4 Equilibrium diagram of the beam element

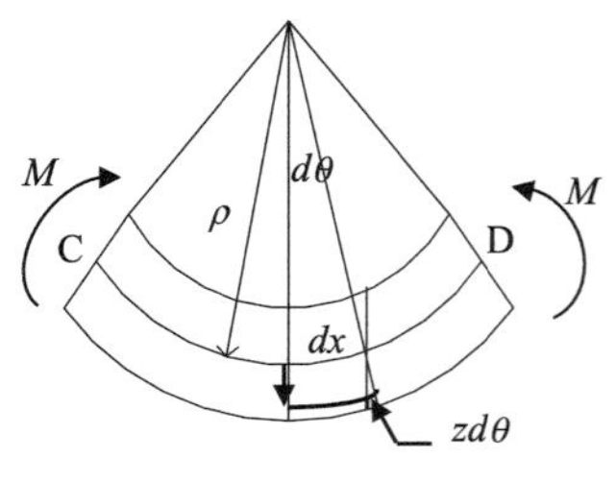

Fig. 6.5 Deformation of element under pure bending

The stress is

$$\sigma_{xx} = E\varepsilon_{xx} = \frac{Ez}{\rho} \tag{6.22}$$

As before let the beam be of width b, thickness t. Force on element (see Fig. 6.2) at distance z thickness dz and width b is

$$dF = \sigma_{xx} b dz = E\varepsilon_{xx} b dz = \frac{Ebz}{\rho} dz \tag{6.23}$$

Taking moment of this elemental force and integrating over the entire thickness from $-1/2\ t$ to $+1/2\ t$, we get the familiar bending moment of the beam.

$$M_{yy} = \int_{-\frac{1}{2}t}^{+\frac{1}{2}t} \frac{Eb}{\rho} z^2 dz = \frac{EI}{\rho} \tag{6.24}$$

The curvature $1/\rho$ is proportional to the bending moment and inversely proportional to EI is the flexural rigidity.

Comparing with Eq. (6.21), we find the beam curvature is

$$\frac{1}{\rho} = \frac{d^2w}{dx^2} = \frac{M}{EI} \tag{6.25}$$

We can also write the stress and strain in terms of bending moment

$$\begin{aligned} \sigma_{xx} &= \frac{Ez}{\rho} = \frac{M}{I} z \\ \varepsilon_{xx} &= \frac{z}{\rho} = \frac{M}{EI} z \end{aligned} \tag{6.26}$$

The slope and deflection relations of the beam are

$$\begin{aligned} \frac{dw}{dx} &= \theta \\ \frac{d^2w}{dx^2} &= \frac{d\theta}{dx} = \frac{1}{\rho} = \frac{M}{EI} \end{aligned} \tag{6.27}$$

Noticing that the slope decreases when the curvature is concave upwards writing the moment equation and differentiating further, we get an ordinary linear constant coefficient differential equation.

$$\begin{aligned} EI\frac{d^2w}{dx^2} &= -M \\ EI\frac{d^3w}{dx^3} &= -\frac{dM}{dx} = -V \\ EI\frac{d^4w}{dx^4} &= -\frac{dV}{dx} = q \end{aligned} \tag{6.28}$$

This equation is solved with appropriate boundary conditions to obtain the deflections. An important aspect on boundary conditions in the Strength of Materials approach is the fact that one has to look at the supports and identify what would be the displacements, slopes, shear forces and bending moments here, whereas with the energy approach, these boundary conditions are a part and parcel of the total physics and its derivation and therefore the solution should take into account the derived kinematic and natural boundary conditions. The governing differential equations in (6.11) and above in (6.28) are however same. The strength of Materials approach is given below.

- Idealize the structure by a beam model with variable cross-section. From the load pattern determine Shear Force along the neutral axis by numerical or graphical integration and plot.

$$\begin{aligned} &\text{Load} \rightarrow q(x) \\ &\frac{dV_x}{dx} = -q \rightarrow \text{Integrate to get } V_x = \int q dx \end{aligned}$$

- Integrate numerically or graphically and get the bending moment diagram.

$$\frac{dM_x}{dx} = V_x \rightarrow M_x = \int V_x dx$$

- Determine area moment of inertia I and obtain bending stress and strain.

$$\sigma_{xx} = \frac{M}{I}z; \quad \varepsilon_{xx} = \frac{M}{EI}z$$

- Divide bending moment M by EI integrate and determine slope.

$$\frac{d\theta}{dx} = \frac{M}{EI} \rightarrow \theta = \int \frac{M}{EI} dx$$

- Integrate numerically or graphically the slope to obtain the deflection curve.

$$\frac{dw}{dx} = \theta \rightarrow w = \int \theta dx$$

Stodola and Viannello developed a graphical method of integration of the differential equation in 1910 to design steam turbine shafts whose cross-section varies in steps based on the design requirements. Their industrial design is illustrated in Fig. 6.6.

In the absence of modern computational facilities, this was the best method adopted and engineers were trained to use T-Squares and Set-Squares on a drafting table. Obviously this requires highly skilled draftsmen, who do not exist anymore because of commercial drafting tools using computers. The time is a major factor in this highly competitive world and therefore these graphical methods have gone into history.

Macaulay in 1919 developed a numerical integration method rather than a graphical procedure used by Stodola. We illustrate Macaulay method as follows by considering a simply supported beam.

Let all the loads be considered as point loads; determine the support reactions R_1 and R_2. Taking a section at distance x that includes all the loads to its left including the left support reaction R_1 as in Fig. 6.7 write the bending moment and integrate as follows.

$$EI\frac{d^2y}{dx^2} = R_1[x] - F_1[x-a] - F_2[x-b] - F_3[x-c]$$

$$EI\frac{dy}{dx} = R_1\frac{[x]^2}{2} - F_1\frac{1}{2}[x-a]^2 - F_2\frac{1}{2}[x-b]^2 - F_3\frac{1}{2}[x-c]^2 + A$$

$$EIy = R_1\frac{[x]^3}{6} - F_1\frac{1}{6}[x-a]^3 - F_2\frac{1}{6}[x-b]^3 - F_3\frac{1}{6}[x-c]^3 + Ax + B$$

Use boundary conditions, solve for A and B.

Worked Example 6.6

Consider the case of a beam 6 m long with two point loads 30,000 N equidistant and symmetrically placed between the supports. The flexural stiffness is 200 M Nm^2. Find the deflection at the middle length of the beam.

The two support reactions are equal and 30,000 N each.

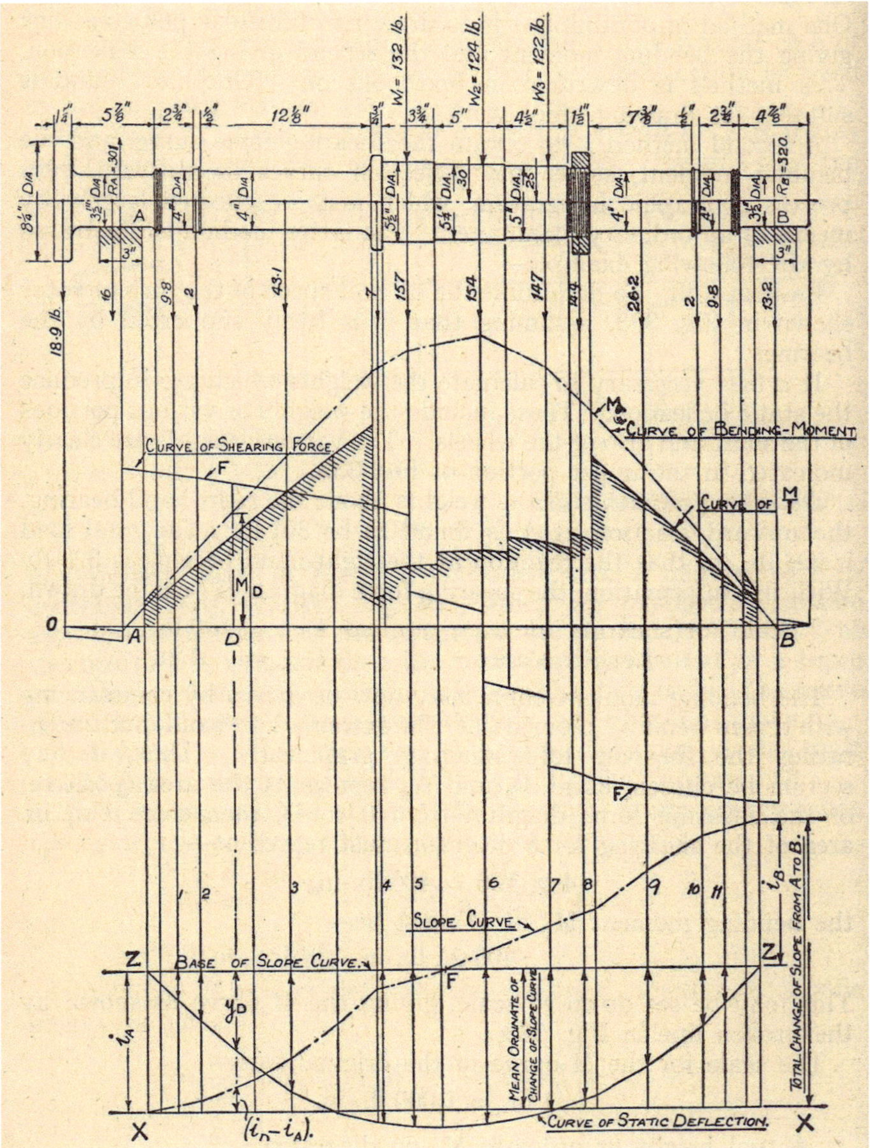

Fig. 6.6 Graphical method of integration in finding deflections

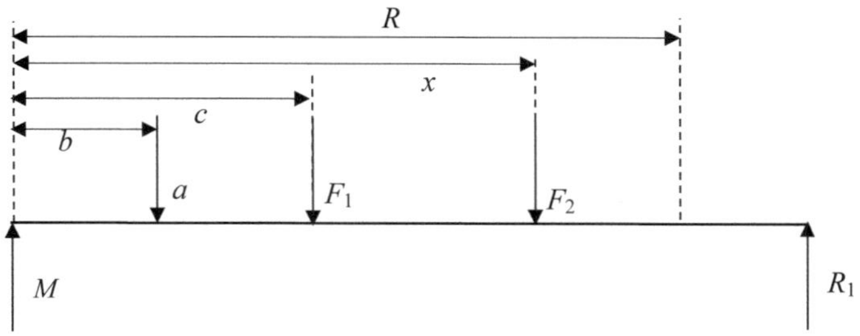

Fig. 6.7 Beam with point loads—Macaulay's method

$$EI\frac{d^2y}{dx^2} = 30{,}000[x] - 30{,}000[x-2] - 30{,}000[x-4]$$

$$EI\frac{dy}{dx} = 30{,}000\frac{[x]^2}{2} - 30{,}000\frac{1}{2}[x-2]^2 - 30{,}000\frac{1}{2}[x-4]^2 + A$$

$$EIy = 30{,}000\frac{[x]^3}{6} - 30{,}000\frac{1}{6}[x-2]^3 - 30{,}000\frac{1}{6}[x-4]^3 + A[x] + B$$

At $x = 0$, $y = 0$, ignoring any bracket with negative value, we have $B = 0$.
At $x = 6$, $y = 0$, therefore

$$0 = 30{,}000\frac{[6]^3}{6} - 30{,}000\frac{1}{6}[4]^3 - 30{,}000\frac{1}{6}[2]^3 + A[6] \Rightarrow$$

$$A = -\frac{30{,}000}{36}(216 - 64 - 8) = -120{,}000$$

$$\therefore y = \frac{1}{EI}\left(30{,}000\frac{[x]^3}{6} - 30{,}000\frac{1}{6}[x-2]^3 - 30{,}000\frac{1}{6}[x-4]^3 - 120{,}000[x]\right)$$

$$y|_{x=3} = \frac{1}{200\times 10^6}\left(30{,}000\frac{3^3}{6} - 30{,}000\frac{1}{6}[1]^3 - 120{,}000[3]\right)$$

$$= \frac{3}{20{,}000}\left(\frac{3^3}{6} - \frac{1}{6}1^3 - 12\right) = \frac{3}{6\times 20{,}000}(27 - 1 - 72)$$

$$= -\frac{3\times 46}{6\times 20{,}000}\,\mathrm{m} = -1.15\ \mathrm{mm}$$

Worked Example 6.7

A cantilever of length l and area of cross-section A is fixed with deflection and slope zero at left end and free at the right end. The Young's modulus of the material is E. The second moment of area of the cross-section about yy axis is determined to be I_{yy}.

The bending moment at distance x from the left end is $-\frac{1}{2}q(l-x)^2$. From Eq. (6.28)

$$EI\frac{d^2w}{dx^2} = -M = \frac{1}{2}q(l-x)^2 \tag{a}$$

The above can be simplified

$$\frac{d^2w}{dx^2} = \frac{q}{2EI}\left(l^2 - 2lx + x^2\right) \tag{b}$$

Integrating once

$$\frac{dw}{dx} = \frac{q}{2EI}\left(l^2x - lx^2 + \frac{1}{3}x^3\right) + A \tag{c}$$

At $x = 0$ the slope $dw/dx = 0$, therefore $A = 0$

$$\therefore \frac{dw}{dx} = \frac{q}{2EI}\left(l^2x - lx^2 + \frac{1}{3}x^3\right) \tag{d}$$

Integrating once more

$$w = \frac{q}{2EI}\left(\frac{1}{2}l^2x^2 - \frac{1}{3}lx^3 + \frac{1}{12}x^4\right) + B \tag{e}$$

At $x = 0$ the deflection $w = 0$, Therefore $B = 0$

$$w = \frac{q}{2EI}\left(\frac{1}{2}l^2x^2 - \frac{1}{3}lx^3 + \frac{1}{12}x^4\right) \tag{f}$$

At $x = l$ the tip deflection w is

$$w_{x=l} = \frac{ql^4}{2EI}\left(\frac{1}{2} - \frac{1}{3} + \frac{1}{12}\right) = \frac{ql^4}{8EI} \tag{g}$$

6.4 Beam Solution by Energy Method

Consider a simple example of a cantilever with uniformly distributed load. Let's approximate the deflection $w(x)$ with x non-dimensionalized to vary from 0 to 1 satisfying boundary conditions by a polynomial of fourth degree where a is a constant to be determined.

Let the deflection be written as a polynomial of 4th degree to account for four boundary conditions.

$$w(X) = a_0 + a_1X + a_2X^2 + a_3X^3 + a_4X^4 \tag{6.29}$$

where $X = x/l$. Since w is zero at $X = 0$, $a_0 = 0$. Then

$$\begin{aligned} w(X) &= a_1X + a_2X^2 + a_3X^3 + a_4X^4 \\ \frac{dw(X)}{dX} &= a_1 + 2a_2X + 3a_3X^2 + 4a_4X^3 \end{aligned} \tag{6.30}$$

Now w' is zero at $X = 0$, $a_1 = 0$, therefore

$$\begin{aligned} w(X) &= a_2X^2 + a_3X^3 + a_4X^4 \\ \frac{dw(X)}{dX} &= 2a_2X + 3a_3X^2 + 4a_4X^3 \\ \frac{d^2w(X)}{dX^2} &= 2a_2 + 6a_3X + 12a_4X^2 \end{aligned} \tag{6.31}$$

At the free end $X = 1$, the bending moment is zero, therefore

$$\frac{d^2w(X)}{dX^2} = 2a_2 + 6a_3 + 12a_4 = 0 \tag{6.32}$$

Differentiating once more

$$\frac{d^3w(X)}{dX^3} = 6a_3 + 24a_4X \tag{6.33}$$

At the free end $X = 1$, the shear force is zero, therefore

$$\frac{d^3w(X)}{dX^3} = 6a_3 + 24a_4 = 0 \Rightarrow a_4 = -\frac{1}{4}a_3 \tag{6.34}$$

Substituting in $2a_2 + 6a_3 + 12a_4 = 0$ we get

$$\begin{aligned} 2a_2 + 6a_3 - 3a_3 &= 0 \\ a_2 &= -\frac{3}{2}a_3 \end{aligned} \tag{6.35}$$

We can now eliminate a_2 and a_4 in (6.31)

$$\begin{aligned} w(X) &= -\frac{3}{2}a_3X^2 + a_3X^3 - \frac{1}{4}a_3X^4 \\ &= a_3\left(-\frac{3}{2}X^2 + X^3 - \frac{1}{4}X^4\right) \end{aligned}$$

Multiply by 4 and redefining the constant as a, the deflection assumed is written as

$$\begin{aligned} w(X) &= a\left(6X^2 - 4X^3 + X^4\right) \\ \frac{d^2w}{dX^2} &= \frac{a}{L^2}\left(12 - 24X + 12X^2\right) \end{aligned} \tag{6.36}$$

Total potential energy π is

$$\begin{aligned} \pi &= \frac{1}{2}\int_0^L EI\left(\frac{d^2w}{dx^2}\right)^2 dx - \int_0^L q(x)w(x)dx \\ &= \frac{1}{2}\frac{EI}{L^3}\int_0^1 \left(\frac{d^2w}{dX^2}\right)^2 dX - L\int_0^1 q(X)a\left(6X^2 - 4X^3 + X^4\right)dX \\ &= \frac{1}{2}\frac{EI}{L^3}\int_0^1 \left[a\left(12 - 24X + 12X^2\right)\right]^2 dX - L\int_0^1 q(X)a\left(6X^2 - 4X^3 + X^4\right)dX \end{aligned} \tag{6.37}$$

Minimize the total potential energy by making its derivative with respect to a to be zero

$$\frac{\partial\pi}{\partial a} = a\frac{EI}{L^3}\int_0^1 \left[\left(12 - 24X + 12X^2\right)\right]^2 dX - L\int_0^1 q(X)\left(6X^2 - 4X^3 + X^4\right)dX = 0 \tag{6.38}$$

The first Integral in the above is

$$\begin{aligned} \int_0^1 \left(12 - 24X + 12X^2\right)^2 dX &= \int_0^1 \left(144 - 576X + 864X^2 - 576X^3 + 144X^4\right)dX \\ &= 144\left(X - \frac{4}{2}X^2 + \frac{6}{3}X^3 - \frac{4}{4}X^4 + \frac{1}{5}X^5\right)_0^1 \\ &= 144\left(1 - \frac{4}{2} + \frac{6}{3} - \frac{4}{4} + \frac{1}{5}\right) = \frac{144}{5} \end{aligned} \tag{6.39a}$$

Likewise the second integral is

$$\begin{aligned} \int_{04}^1 \left(6X^2 - 4X^3 + X^4\right)dX &= \left(\frac{6}{3}X^3 - \frac{4}{4}X^4 + \frac{1}{5}X^5\right)_0^1 \\ &= \left(\frac{6}{3} - \frac{4}{4} + \frac{1}{5}\right) = \frac{6}{5} \end{aligned} \tag{6.39b}$$

Substituting in (6.38)

$$\frac{144}{5}\frac{EI}{L^3}a = \frac{6}{5}qL \quad \therefore a = \frac{1}{24}\frac{L^4}{EI}q \tag{6.40}$$

Therefore the deflection as a function of x is given by

$$w(x) = \frac{qL^4}{24EI}\left(6X^2 - 4X^3 + X^4\right) \tag{6.41}$$

The tip deflection is obtained by substituting $X = 1$

$$\begin{aligned} w_{\text{tip}} &= \frac{qL^4}{24EI}\left(6X^2 - 4X^3 + X^4\right) \\ &= \frac{qL^4}{24EI}(6 - 4 + 1) = \frac{qL^4}{8EI} \end{aligned} \tag{6.42}$$

The above is a correct value, but the function in (6.41) is not necessarily the exact solution. The approximate solution is lower (always) bound because when we assumed a solution other than the exact one, we have bent the beam by additional energy to take it to an altered position from the exact one; this means we have stiffened the beam and therefore lower approximate solution.

We can adopt two ways of getting a better approximate solution. One way of improving the accuracy is to use more number of terms, each term increasing in degree but still satisfying the boundary conditions. The error then gets reduced gradually or we say that the solution converges.

Such were the methods developed in 1960s. One may call them single element models. Main problem here is the ability to capture the intricate geometry around the cut-outs …

In the finite element method we use more sub domains (finite elements) and satisfy the assumed solution in all these elements thus improving the accuracy or getting a converged solution closer to the exact one.

Remember in either case, we are avoiding complex theory of elasticity on one side and not restricting to beam strength of material theory on the other side. That is how engineers love these solutions to get realistic good solutions for practical structures.

It is important to remember that basic energy principle is used in the derivation of finite element method; that is why energy principles and theory of elasticity are important. Also note the contrast: This is finite element as against familiar infinitesimally small element used in calculus. Each element is a structure and should satisfy the governing equations or theory of elasticity as assumed in the derivation of the elements.

We have seen two approaches for a beam problem. In the Theory of Elasticity approach, we used an approximate displacement field consistent with any assumptions (technical theory of beams), derived strain and stress fields in terms of the displacements and arrived at a differential equation to solve the displacements first and then stresses and strains. This is more mathematical in approach which engineers hate; however has a great potential to understand complex problems and develop numerical methods suitable for high speed computer applications, particularly the finite element method that can address the needs of practical real life complex engineering structures. We will discuss this later.

In the Strength of Materials approach, we used an equilibrium diagram approach; defined shear force and bending moment as a function of space coordinate (s). Curvature, stress and strain are determined without determining the displacement field, thus avoiding complex mathematics. The deflection is obtained by solving an ordinary differential equation or by integration of the bending moment diagram. Main advantage is that the approach is simple and the stress field can be determined quickly for a simple structure. We use this advantage to determine average stress field in a complex structure by simplifying it and ensure that the finite element solutions are reasonably correct; so-called *P/A* approach.

Both these approaches are extensively used until recent times. Engineers developed graphical methods over a century ago and extensively practiced till 1950s. As graphical methods are tedious and need highly skilled labor, tabular methods were developed to do this integration, particularly when we need to do dynamics of beams. These numerical techniques were fore runners for computerization.

Finite element method was evolved 50 years ago to address the demands of complex structures with intricate geometry and determine peak stresses accurately as against global beam (or plate …) stresses in an idealized structure. The idealized structures yield only average values across cross-sections but fail to determine peak stresses at stress raiser locations and thus Factors of Safety (actually factors of ignorance) are extensively used in design. Finite element method is able to remove this limitation in designs; thanks to high speed computation that exists today.

6.5 Beam Finite Element

Discretize the given beam into several finite elements with one element shown in Fig. 6.8 in local coordinates marked by a hat. The element has two nodes 1 and 2. The nodal deflections are measured positive in the direction of the coordinate axis. The nodal slopes are positive in counterclockwise direction according to the right hand rule and similarly the bending moments and shear forces are as shown.

Assume the transverse displacement to be a cubic polynomial which satisfies the governing differential equation of a beam. In addition, we also note that the cubic displacement shape function satisfies the continuity condition of both the deflection and slope at the nodes.

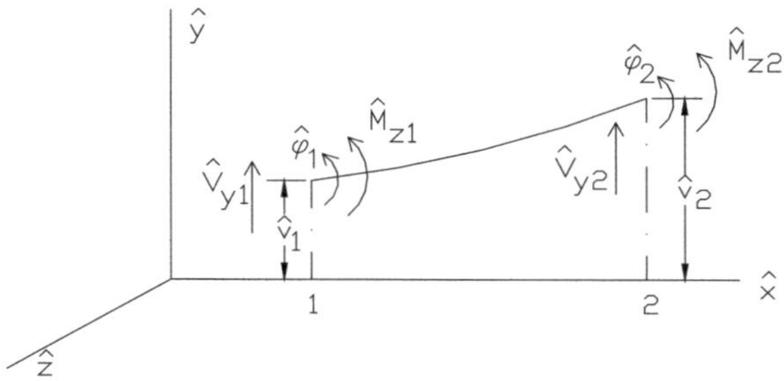

Fig. 6.8 Beam element in local coordinates

$$\hat{v}(\hat{x}) = a_1\hat{x}^3 + a_2\hat{x}^2 + a_3\hat{x} + a_4 \tag{6.43}$$

With the help of the boundary conditions of the element at the two nodes, we have

$$\begin{aligned} \hat{v}(0) &= \hat{v}_1 = a_4 \\ \frac{d\hat{v}(0)}{d\hat{x}} &= \hat{\phi}_1 = a_3 \\ \hat{v}(L) &= \hat{v}_2 = a_1L^3 + a_2L^2 + a_3L + a_4 \\ \frac{d\hat{v}(L)}{d\hat{x}} &= \hat{\phi}_2 = 3a_1L^2 + 2a_2L + a_3 \end{aligned} \tag{6.44}$$

The shape function can now be written as

$$\begin{aligned} \hat{v}(\hat{x}) = &\left[\frac{2}{L^3}(\hat{v}_1 - \hat{v}_2) + \frac{1}{L^2}\left(\hat{\phi}_1 + \hat{\phi}_2\right)\right]\hat{x}^3 \\ &- \left[\frac{3}{L^2}(\hat{v}_1 - \hat{v}_2) + \frac{1}{L}\left(2\hat{\phi}_1 + \hat{\phi}_2\right)\right]\hat{x}^2 + \hat{\phi}_1\hat{x} + \hat{v}_1 \end{aligned} \tag{6.45}$$

The above equation is expressed for convenience in the form

$$\begin{aligned} \hat{v} &= [N]\{\hat{d}\} \\ &= [N_1 \quad N_2 \quad N_3 \quad N_4]\{\hat{d}\} \end{aligned} \tag{6.46}$$

where d are nodal degrees of freedom and N are shape function given by

$$\{\hat{d}\} = \begin{Bmatrix} \hat{v}_1 \\ \hat{\phi}_1 \\ \hat{v}_2 \\ \hat{\phi}_2 \end{Bmatrix} \tag{6.47}$$

$$
\begin{aligned}
N_1 &= \frac{1}{L^3}\left(2\hat{x}^3 - 3\hat{x}^2L + L^3\right)\\
N_2 &= \frac{1}{L^3}\left(\hat{x}^3L - 2\hat{x}^2L^2 + \hat{x}L^3\right)\\
N_3 &= \frac{1}{L^3}\left(-2\hat{x}^3 + 3\hat{x}^2L\right)\\
N_4 &= \frac{1}{L^3}\left(\hat{x}^3L - \hat{x}^2L^2\right)
\end{aligned}
\tag{6.48}
$$

The bending moment is now expressed as a function of displacement given by

$$
\begin{aligned}
\hat{M}_z(\hat{x}) &= EI_{zz}\frac{d^2\hat{v}}{d\hat{x}^2}\\
\hat{M}_z(\hat{x}) &= EI_{zz}[B]\{\hat{d}\}
\end{aligned}
\tag{6.49}
$$

where

$$
[B] = \left[N_{1,\hat{x}\hat{x}} \quad N_{2,\hat{x}\hat{x}} \quad N_{3,\hat{x}\hat{x}} \quad N_{4,\hat{x}\hat{x}}\right] \tag{6.50}
$$

and

$$
\begin{aligned}
N_{1,\hat{x}\hat{x}} &= \frac{1}{L^3}(12\hat{x} - 6L)\\
N_{2,\hat{x}\hat{x}} &= \frac{1}{L^3}\left(6\hat{x}L - 4L^2\right)\\
N_{3,\hat{x}\hat{x}} &= \frac{1}{L^3}(-12\hat{x} + 6L)\\
N_{4,\hat{x}\hat{x}} &= \frac{1}{L^3}\left(6\hat{x}L - 2L^2\right)
\end{aligned}
\tag{6.51}
$$

The strain energy is now expressed as

$$
\begin{aligned}
U &= \frac{1}{2}EI_{zz}\int_0^L \hat{v}^2_{,\hat{x}\hat{x}}d\hat{x}\\
&= \frac{1}{2}EI_{zz}\{\hat{d}\}^T\int_0^L [B]^T[B]\{\hat{d}\}d\hat{x}
\end{aligned}
\tag{6.52}
$$

Using Lagrangian approach

$$
\frac{\partial U}{d\{\hat{d}\}} = EI_{zz}\int_0^L [B]^T[B]\{\hat{d}\}d\hat{x} = F \tag{6.53}
$$

Writing

$$F = [\hat{K}]\{\hat{d}\} \tag{6.54}$$

the stiffness matrix of the beam element can be obtained as

$$[\hat{K}] = EI_{zz}\int_0^L \begin{bmatrix} N_{1,\hat{x}\hat{x}} \\ N_{2,\hat{x}\hat{x}} \\ N_{3,\hat{x}\hat{x}} \\ N_{4,\hat{x}\hat{x}} \end{bmatrix} [N_{1,\hat{x}\hat{x}} \quad N_{2,\hat{x}\hat{x}} \quad N_{3,\hat{x}\hat{x}} \quad N_{4,\hat{x}\hat{x}}]d\hat{x}$$

$$[\hat{K}] = \frac{EI_{zz}}{L^3}\begin{bmatrix} 12 & 6L & -12 & 6L \\ & 4L^2 & -6L & 2L^2 \\ & & 12 & -6L \\ & & & 4L^2 \end{bmatrix}_{\text{sym}} \tag{6.55}$$

Worked Example 6.8

Consider a cantilever beam represented as one element. Node 1 is taken at fixed end and node 2 at free end. There is just one tip load P. The local and global coordinates in this case are same. Equation (6.54) gives

$$[K]\{d\} = \{F\}$$

$$\frac{EI}{L^3}\begin{bmatrix} 12 & 6L & -12 & 6L \\ & 4L^2 & -6L & 2L^2 \\ & & 12 & -6L \\ & & & 4L^2 \end{bmatrix}\begin{Bmatrix} v_1 \\ \phi_1 \\ v_2 \\ \phi_2 \end{Bmatrix} = \begin{Bmatrix} 0 \\ 0 \\ P \\ 0 \end{Bmatrix} \tag{6.56}$$

Applying boundary conditions, we get

$$[K]\{d\} = \{F\}$$

$$\frac{EI}{L^3}\begin{bmatrix} 12 & 6L & -12 & 6L \\ & 4L^2 & -6L & 2L^2 \\ & & 12 & -6L \\ & & & 4L^2 \end{bmatrix}\begin{Bmatrix} 0 \\ 0 \\ v_2 \\ \varphi_2 \end{Bmatrix} = \begin{Bmatrix} 0 \\ 0 \\ P \\ 0 \end{Bmatrix} \tag{6.57}$$

$$\frac{EI}{L^3}\begin{bmatrix} 12 & -6L \\ & 4L^2 \end{bmatrix}\begin{Bmatrix} v_2 \\ \phi_2 \end{Bmatrix} = \begin{Bmatrix} P \\ 0 \end{Bmatrix}$$

The above gives two coupled linear algebraic equations in deflection and slope at free end.

$$\begin{aligned}\frac{12EI}{L^3}v_2 - \frac{6EI}{L^2}\phi_2 &= P\\ -\frac{6EI}{L^2}v_2 + \frac{4EI}{L}\phi_2 &= 0\end{aligned} \tag{6.58}$$

Solving

$$\begin{aligned}\phi_2 &= \frac{3}{2L}v_2\\ v_2 &= P\frac{L^3}{3EI}\end{aligned} \tag{6.59}$$

To understand taking more elements, consider a beam fixed at both the ends, Let us divide the fixed-fixed beam into two elements (which we can still solve by hand to understand the process). Write the two elemental stiffness matrices and assemble them in global matrix considering all global degrees of freedom. Then

$$[\hat{K}] = \frac{EI_{zz}}{L^3}\begin{matrix}[v_1 & \phi_1 & v_2 & \phi_2 & v_3 & \phi_3]\end{matrix}\begin{bmatrix}12 & 6L & -12 & 6L & & \\ & 4L^2 & -6L & 2L^2 & & \\ & & 24 & 0 & -12 & 6L\\ & & & 8L^2 & -6L & 2L^2\\ & & & & 12 & -6L\\ & & & & & 4L^2\end{bmatrix} \tag{6.60}$$

Let the beam be loaded at mid-point node 2 and apply boundary conditions at both the fixed ends 1 and 3 nodes to give

$$\frac{EI}{L^3}\begin{bmatrix}24 & 0\\ 0 & 8L^2\end{bmatrix}\begin{Bmatrix}v_2\\ \phi_2\end{Bmatrix} = \begin{Bmatrix}P\\ 0\end{Bmatrix} \tag{6.61}$$

This gives

$$\begin{aligned}&\frac{24EI}{L^3}v_2 - 0\phi_2 = P\\ &0v_2 + \frac{8EI}{L}\phi_2 = 0 \rightarrow \phi_2 = 0\\ &v_2 = P\frac{L^3}{24EI}\end{aligned} \tag{6.62}$$

Worked Example 6.9

Consider a rectangular square cross-section 0.1 × 0.1 m cantilever beam of length 1 m as shown in Fig. Worked Example 6.9.1. It is clamped on the left side and has a

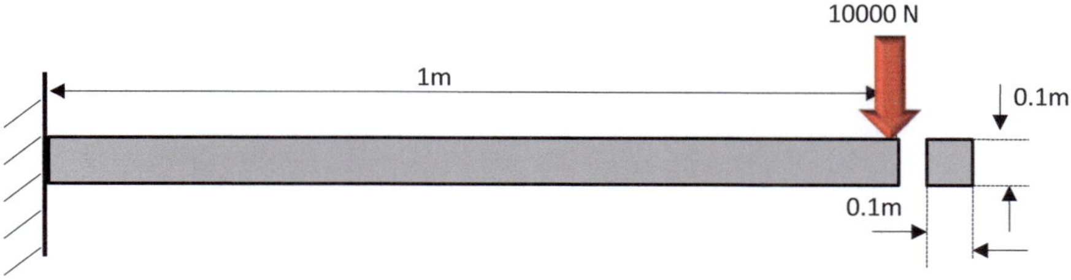

Fig. Worked Example 6.9.1 Cantilever beam

point force of 10 kN acting downward on the right end of the beam. The beam is made of steel with Young's Modulus 200 GPa.

From theory at $x = 1$ m

$$u = \frac{PL}{EI}\left(\frac{x^2}{2} - \frac{x^3}{6L}\right) = \frac{10{,}000 \times 12}{200 \times 10^9 \times 0.1^4}\left(\frac{1}{2} - \frac{1}{6}\right) = 2 \text{ mm}$$

To carry out the analysis by FEM, the beam is modeled with two elements as shown in Fig. Worked Example 6.9.2.

The deformed beam obtained is given in Fig. Worked Example 6.9.3.

The bending deflection at the free end is upper bound 2.0154 mm as compared to the closed form solution result $u = \frac{PL}{EI}\left(\frac{x^2}{2} - \frac{x^3}{6L}\right) = 2$ mm.

Also the deflection at the mid-point $u|_{x=\frac{1}{2}L} = \frac{10{,}000 \times 12}{200 \times 10^9 \times 0.1^4}\left(\frac{1}{8} - \frac{1}{48}\right) = 0.625$ mm is in good agreement with the FE result.

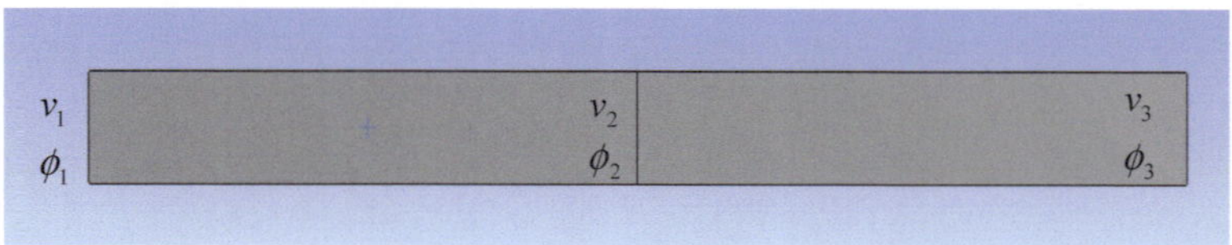

Fig. Worked Example 6.9.2 Beam model with two elements

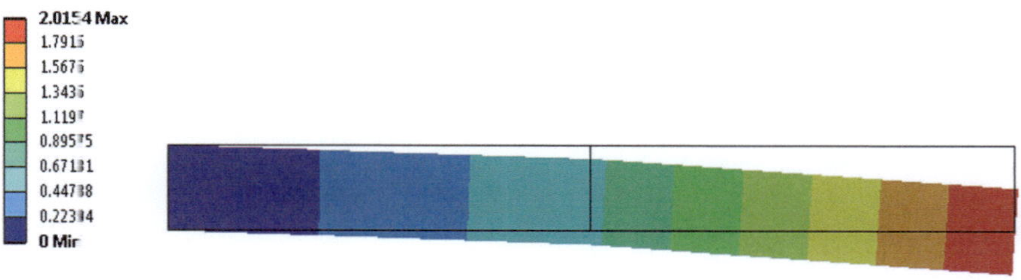

Fig. Worked Example 6.9.3 Deformation of beam with two elements

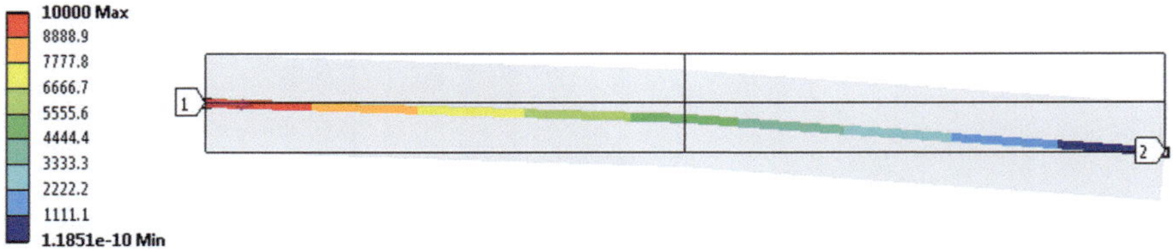

Fig. Worked Example 6.9.4 Bending moment with two elements

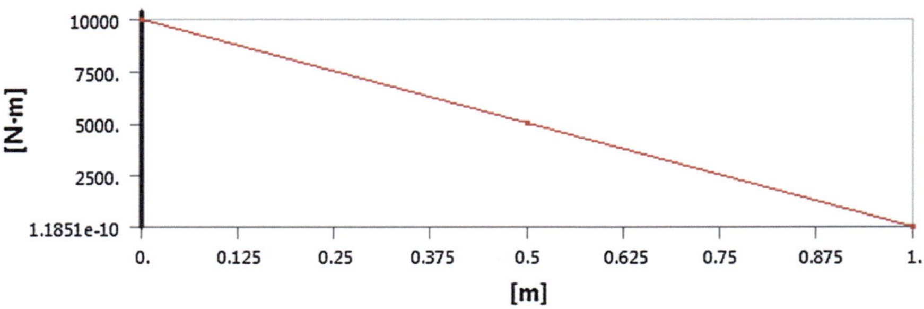

Fig. Worked Example 6.9.5 Bending moment along the beam

At the fixed end, the bending moment is $M = F \times L = 10{,}000 \times 1 = 10{,}000$ Nm.

The bending moment obtained from FEM with two elements is shown in Fig. Worked Example 6.9.4.

A plot of Bending Moments is given in Fig. Worked Example 6.9.5.

The maximum bending stress using Euler-Bernoulli beam theory is given by

$$\sigma_{\max} = \frac{My}{I} = \frac{10{,}000 \times 0.05}{\frac{0.1 \times 0.1^3}{12}} = 60 \text{ MPa}$$

From FEM, the bending stress obtained is given in Fig. Worked Example 6.9.6. Note that this is a one dimensional model and that the stress shown is maximum stress along the axis.

The above are also average stress over the length of the beam. Using only two elements, these yields a minimum value of 7.11×10^{-13} Pa for the maximum bending stress. This is a small error created by the small number of elements used. The minimum value of bending stress should be zero because there is no moment at the free end. The maximum value of stress at the fixed end is in agreement with the theoretical value.

The accuracy can be improved by increasing the number of elements in FE model. Figure Worked Example 6.9.7 shows the Beam Model with Ten Elements.

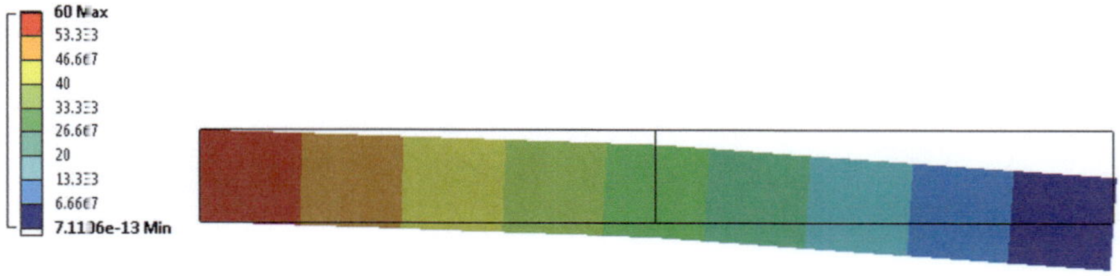

Fig. Worked Example 6.9.6 Bending stress along the axis

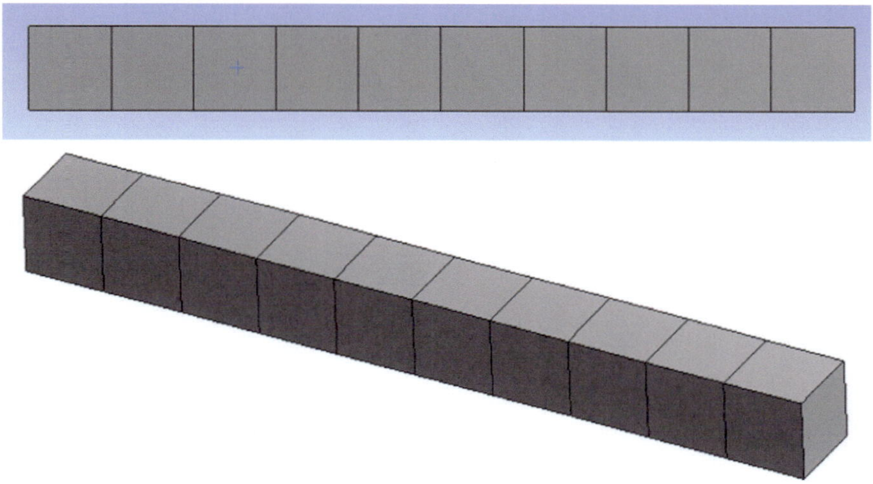

Fig. Worked Example 6.9.7 Beam model with ten elements

Table Worked Example 6.9 Convergence study

	Tip deformation (m)	Max bending stress (MPa)
Theory values	0.002	60
1 element FEA	0.002015447	60
2 element FEA	0.002015440	60
10 element FEA	0.002015437	60

As one can see from the Table Worked Example 6.9 the results do not change as the mesh is refined. The reason that the results do not change is that they have already converged. The exact solution for the deformation is cubic and element used employs cubic interpolation. Thus, for the simple cantilever beam setup the numerical method converges very quickly.

6.5.1 Additional Stiffness from Axial Load

In Chap. 2 we dealt with axially loaded columns without any bending. Usually there is some bending as the section above present that makes the 1-D structure a beam column.

Figure 6.3 shows the effect of axial compressive force; its effect is to decrease the strain energy by $\frac{1}{2}S\int \widehat{v}_{,x}^{2}\, d\widehat{x}$. The contribution of axial compressive force to the strain energy of the element U_S from Eq. (6.12) is

$$U_S = -\frac{1}{2}S\int \widehat{v}_{,x}^{2}\, d\widehat{x} \tag{6.63}$$

From (6.46)

$$\begin{aligned} \hat{v}_{,x} &= [N_{,x}]\{\hat{d}\} \\ &= [N_{1,x} \quad N_{2,x} \quad N_{3,x} \quad N_{4,x}]\{\hat{d}\} \end{aligned} \tag{6.64}$$

Substituting in (6.43), the additional strain energy to the element is

$$U_S = -\frac{1}{2}S\{\widehat{d}\}^T\int [N_{,x}]^T[N_{,x}]\{\widehat{d}\}d\widehat{x} \tag{6.65}$$

Then

$$\frac{\partial U_S}{d\{\widehat{d}\}} = -S\int [N_{,x}]^T[N_{,x}]\{\widehat{d}\}d\widehat{x} \tag{6.66}$$

With the help of (6.48)

$$\begin{aligned} N_1 &= \frac{1}{L^3}\left(2\hat{x}^3 - 3\hat{x}^2L + L^3\right) \\ N_2 &= \frac{1}{L^3}\left(\hat{x}^3L - 2\hat{x}^2L^2 + \hat{x}L^3\right) \\ N_3 &= \frac{1}{L^3}\left(-2\hat{x}^3 + 3\hat{x}^2L\right) \\ N_4 &= \frac{1}{L^3}\left(\hat{x}^3L - \hat{x}^2L^2\right) \end{aligned} \tag{6.48}$$

(6.66) gives the additional stiffness given by

$$\left[\hat{K}\right]_S = -\frac{S}{L^6}\int_0^L \left[\left(6\hat{x}^2 - 6\hat{x}L\right) \quad \left(3\hat{x}^2L - 4\hat{x}L^2 + L^3\right) \quad \left(-6\hat{x}^2 + 6\hat{x}L\right) \quad \left(3\hat{x}^2L - 2\hat{x}L\right)\right] \begin{bmatrix} \left(6\hat{x}^2 - 6\hat{x}L\right) \\ \left(3\hat{x}^2L - 4\hat{x}L^2 + L^3\right) \\ \left(-6\hat{x}^2 + 6\hat{x}L\right) \\ \left(3\hat{x}^2L - 2\hat{x}L\right) \end{bmatrix} d\hat{x}$$

$$= -\frac{S}{30L}\begin{bmatrix} 36 & 3L & -36 & 3L \\ & 4L^2 & -3L & -L^2 \\ & & 36 & -3L \\ & & & 4L^2 \end{bmatrix} \tag{6.67}$$

Adding to the beam element stiffness in (6.55), the total beam-column stiffness is

$$\left[\hat{K}\right] = \frac{EI_{zz}}{L^3}\begin{bmatrix} 12 & 6L & -12 & 6L \\ & 4L^2 & -6L & 2L^2 \\ & & 12 & -6L \\ & & & 4L^2 \end{bmatrix} - \frac{S}{30L}\begin{bmatrix} 36 & 3L & -36 & 3L \\ & 4L^2 & -3L & -L^2 \\ & & 36 & -3L \\ & & & 4L^2 \end{bmatrix} \tag{6.68}$$

Consider the cantilever example again with an axial compressive load S, we have

$$[K]\{d\} = \left[\frac{EI_{zz}}{L^3}\begin{bmatrix} 12 & 6L & -12 & 6L \\ & 4L^2 & -6L & 2L^2 \\ & & 12 & -6L \\ & & & 4L^2 \end{bmatrix} - \frac{S}{30L}\begin{bmatrix} 36 & 3L & -36 & 3L \\ & 4L^2 & -3L & -L^2 \\ & & 36 & -3L \\ & & & 4L^2 \end{bmatrix}\right]\begin{Bmatrix} 0 \\ 0 \\ v_2 \\ \phi_2 \end{Bmatrix} = \begin{Bmatrix} 0 \\ 0 \\ P \\ 0 \end{Bmatrix} \tag{6.69}$$

Simplifying

$$\left[\frac{EI_{zz}}{L^3}\begin{bmatrix} 12 & -6L \\ & 4L^2 \end{bmatrix} - \frac{S}{30L}\begin{bmatrix} 36 & -3L \\ & 4L^2 \end{bmatrix}\right]\begin{Bmatrix} v_2 \\ \phi_2 \end{Bmatrix} = \begin{Bmatrix} P \\ 0 \end{Bmatrix} \tag{6.69a}$$

$$\begin{aligned}&\left\{12\frac{EI_{zz}}{L^3}-\frac{36S}{30L}\right\}v_2+\left\{-\frac{6EI_{zz}}{L^2}+\frac{S}{10}\right\}\phi_2=P\\&\left\{-6\frac{EI_{zz}}{L^2}+\frac{S}{10}\right\}v_2+\left\{4\frac{EI_{zz}}{L}-\frac{4SL}{30}\right\}\phi_2=0\end{aligned}\tag{6.69b}$$

From the second equation in the above, we have

$$\phi_2=\frac{6\frac{EI_{zz}}{L^2}-\frac{S}{10}}{4\frac{EI_{zz}}{L}-\frac{4SL}{30}}v_2\tag{6.69c}$$

Substituting for ϕ_2 from the above

$$v_2=P\frac{\left(4\frac{EI_{zz}}{L}-\frac{4SL}{30}\right)}{\left\{12\frac{EI_{zz}}{L^3}-\frac{36S}{30L}\right\}\left\{4\frac{EI_{zz}}{L}-\frac{4SL}{30}\right\}+\left\{\frac{6EI_{zz}}{L^2}-\frac{S}{10}\right\}^2}\tag{6.70}$$

From (6.17) we defined Euler Buckling Load to be $S_{\text{Euler}}=EI\left(\frac{\pi}{L}\right)^2$.

$$\begin{aligned}v_2&=\frac{P}{EI}\frac{\left(\frac{4}{L}-\frac{4}{30L}\pi^2\right)}{\left\{12\frac{1}{L^3}-\frac{36}{30L^3}\pi^2\right\}\left\{\frac{4}{L}-\frac{4}{30L}\pi^2\right\}+\left\{\frac{6}{L^2}-\frac{1}{10}\left(\frac{\pi}{L}\right)^2\right\}^2}\\&=\frac{P}{EI}\frac{\frac{1}{L}\left(4-\frac{4}{30}\pi^2\right)}{\left\{12\frac{1}{L^4}-\frac{36}{30L^4}\pi^2\right\}\left\{\left(4-\frac{4}{30}\pi^2\right)\right\}+\left\{\frac{6}{L^2}-\frac{1}{10}\left(\frac{\pi}{L}\right)^2\right\}^2}\end{aligned}$$

Dividing the numerator and denominator by $\left(4-\frac{4}{30}\pi^2\right)$

$$v_2=\frac{PL^3}{EI}\frac{1}{\left[\left\{12-\frac{6}{5}\pi^2\right\}+\frac{\left\{6-\frac{1}{10}\pi^2\right\}^2}{\left(4-\frac{4}{30}\pi^2\right)}\right]}\approx\frac{8}{75}\frac{PL^3}{EI}$$

Beam elements are obviously simple and they are the earliest ones to be developed. Plates have two dimensions and elements are developed to take into account Plate theories and similarly shells. Most useful can be solid elements that can model a given solid with intricate geometry, e.g., Turbine disks and blades. Today you may have an impressive array of elements in any commercial structural codes.

6.6 Buckling Revisited

Consider the homogeneous part of the Eq. (6.15b)

$$\frac{d^2}{dx^2}\left[\left(\frac{d^2w}{dx^2}\right) - \lambda^2 w\right] = 0 \tag{6.71}$$

The general solution for this problem is

$$w(x) = A \sin \lambda x + B \cos \lambda x + Cx + D \tag{6.72}$$

Equation (6.16) $w = \sin\frac{i\pi x}{L}$ is a valid solution only for simply supports ends. Consider the case of fixed-fixed conditions of the beam, i.e.,

$$\begin{aligned} w(x=0) &= w'(x=0) = 0 \\ w(x=L) &= w'(x=L) = 0 \end{aligned} \tag{6.73}$$

The above substituted in (6.72) gives

$$\begin{aligned} w(0) &= 0 \Rightarrow B + D = 0 \\ w'(0) &= 0 \Rightarrow A\lambda + C = 0 \\ w(L) &= 0 \Rightarrow A \sin \lambda L + B \cos \lambda L + CL + D = 0 \\ w'(L) &= 0 \Rightarrow A\lambda \cos \lambda L - B\lambda \sin \lambda L + C = 0 \end{aligned} \tag{6.74}$$

We can write the above in matrix form as

$$\begin{bmatrix} 0 & 1 & 0 & 1 \\ \lambda & 0 & 1 & 0 \\ \sin \lambda L & \cos \lambda L & L & 1 \\ \lambda \cos \lambda L & -\lambda \sin \lambda L & 1 & 0 \end{bmatrix} \begin{Bmatrix} A \\ B \\ C \\ D \end{Bmatrix} = \begin{Bmatrix} 0 \\ 0 \\ 0 \\ 0 \end{Bmatrix} \tag{6.75}$$

To obtain non-trivial solutions to this homogeneous system of equations we must have

$$\begin{vmatrix} 0 & 1 & 0 & 1 \\ \lambda & 0 & 1 & 0 \\ \sin \lambda L & \cos \lambda L & L & 1 \\ \lambda \cos \lambda L & -\lambda \sin \lambda L & 1 & 0 \end{vmatrix} = 0 \tag{6.76}$$

Upon expansion, we have

$$2 - 2\cos\lambda L - \lambda L \sin\lambda L = 0 \tag{6.77}$$

The first root of the above equation is $\lambda L = 2\pi$, i.e.,

$$2\pi = \sqrt{\frac{SL^2}{EI}} \Rightarrow S = 4EI\left(\frac{\pi}{L}\right)^2 \tag{6.78}$$

Recall that for a hinged beam, this load is $S = EI\left(\frac{\pi}{L}\right)^2$.

For any other boundary conditions, we may encounter some difficulties, e.g., a cantilever having an end axial load will create problems in boundary conditions; the nearest we can use is a condition where the deflection is non-zero. The easiest assumption we can make is the shape function to be a half sine wave

$$w = \sin\frac{\pi x}{2L} \tag{6.79}$$

Substituting in (6.71)

$$\begin{aligned}
&\left(\frac{i\pi}{2L}\right)^4 \sin\frac{i\pi x}{2L} + \lambda^2\left(\frac{i\pi}{2L}\right)^2 \sin\frac{i\pi x}{2L} = 0\\
&\left(\frac{\pi}{2L}\right)^2 - \lambda^2 = 0 \Rightarrow \lambda^2 = \left(\frac{\pi}{2L}\right)^2 \Rightarrow S = EI\left(\frac{\pi}{2L}\right)^2
\end{aligned} \tag{6.80}$$

6.7 Galerkin Method for Tapered Beams

Let's take a relook at the derivation of Lagrange equation with 2nd order derivatives in the Functional

$$\begin{aligned}
&\int_{x_1}^{x_2} \left[\frac{d^2}{dx^2}\left(\frac{\partial F}{\partial y''}\right) - \frac{d}{dx}\left(\frac{\partial F}{\partial y'}\right) + \frac{\partial F}{\partial y}\right]\eta dx\\
&\quad - \frac{\partial F}{\partial y'}\eta\bigg|_{x_1}^{x_2} + \frac{\partial F}{\partial y''}\eta'\bigg|_{x_1}^{x_2} - \frac{d}{dx}\left(\frac{\partial F}{\partial y'}\right)\eta\bigg|_{x_1}^{x_2} = 0
\end{aligned} \tag{3.15}$$

In order to satisfy the above equation we have to satisfy

$$\int_{x_1}^{x_2} \left[\frac{d^2}{dx^2}\left(\frac{\partial F}{\partial y''}\right) - \frac{d}{dx}\left(\frac{\partial F}{\partial y'}\right) + \frac{\partial F}{\partial y}\right]\eta dx \tag{3.15a}$$

as well as the definite integrals (boundary conditions)

$$\frac{\partial F}{\partial y'}\eta\Big|_{x_1}^{x_2} + \frac{d}{dx}\left(\frac{\partial F}{\partial y'}\right)\eta\Big|_{x_1}^{x_2} = 0 \tag{3.15b}$$

$$\frac{\partial F}{\partial y''}\eta'\Big|_{x_1}^{x_2} = 0 \tag{3.15c}$$

This led to the differential equation and boundary conditions

$$\frac{d^2}{dx^2}\left(\frac{\partial F}{\partial y''}\right) - \frac{d}{dx}\left(\frac{\partial F}{\partial y'}\right) + \frac{\partial F}{\partial y} = 0 \tag{3.16}$$

subject to Kinematic and Natural boundary conditions as

Kinematic		Natural
y' specified (including set to zero)	or	$-\frac{\partial F}{\partial y''} = 0$
y specified (including set to zero)	or	$-\frac{d}{dx}\left(\frac{\partial F}{\partial y'}\right) - \frac{\partial F}{\partial y'} = 0$

We used (3.16) as the basis to derive the differential equation and made solutions where possible for beams. A differential equation like this cannot unfortunately address tapered beam with varying cross-sections. To be able to do this, let's write (3.15) as

$$\int_{x_1}^{x_2} (Ly)\eta dx + \{(gy)\eta'\}_{x_1}^{x_2} + \{(hy)\eta\}_{x_1}^{x_2} = 0 \tag{6.81}$$

where L, g and h are linear differential operators. When we consider the case that $\eta = 0$, the differential equation symbolically is

$$Ly = 0 \tag{6.82}$$

and we were left with satisfying the boundary conditions, viz.,

$$\{gy = 0\}_{x_1}^{x_2} \text{ and } \{hy = 0\}_{x_1}^{x_2} \tag{6.83}$$

How to address the problem when it is not simple and Eqs. (3.16) or (6.82) are not amenable to yield a solution. Galerkin in 1925 provided an approximate way of solving; it was argued that we need not satisfy $Ly = 0$, instead we can make $\int_{x_1}^{x_2} (L\tilde{y})\eta dx = 0$. Then what η should be used. Galerkin suggests that η can be the assumed solution itself. Let the assumed solution be $\tilde{y}$ and we substitute this in the differential equation, we get an error ε in the differential equation.

$$\varepsilon = L\tilde{y} \tag{6.84}$$

Then

$$\int_{x_1}^{x_2} \varepsilon \tilde{y} dx = 0$$
i.e.,
$$\int_{x_1}^{x_2} (L\tilde{y})\tilde{y} dx = 0 \tag{6.85}$$

This is also called orthogonalization of the error with the assumed solution. Let's consider the following problem. A cantilever beam of length L carries a uniformly distributed load $q(x)$ with rectangular cross-section tapering down from fixed end to free end as shown in Fig. 6.9 defined by the following relations for breadth b and thickness t with $X = \frac{x}{L}$

$$\begin{aligned} b &= b_0(1 - \alpha X) \\ t &= t_0(1 - \beta X) \\ I &= I_0(1 - \alpha X)(1 - \beta X)^3 \end{aligned} \tag{6.86}$$

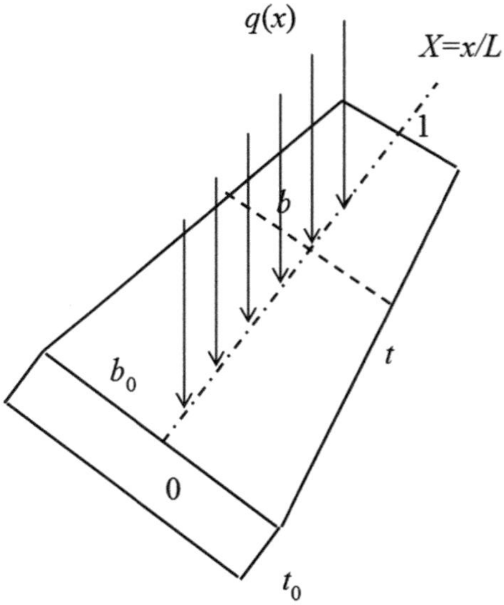

Fig. 6.9 A tapered beam

where $I_0 = \frac{b_0 t_0^3}{12}$

The differential equation is

$$\frac{d^2}{dx^2}\left[EI\left(\frac{d^2w}{dx^2}\right)\right] = q(x)$$

Using (6.76)

$$\frac{d^2}{dX^2}\left[(1-\alpha X)(1-\beta X)^3\frac{d^2w}{dX^2}\right] = \frac{qL^4}{EI_0} \tag{6.87}$$

As before the assumed solution is taken as

$$\begin{aligned} w(X) &= a\left(6X^2 - 4X^3 + X^4\right) \\ \frac{d^2w}{dX^2} &= \frac{a}{L^2}\left(12 - 24X + 12X^2\right) \end{aligned} \tag{6.36}$$

Substituting, the error ε in the differential equation is

$$\varepsilon = \frac{d^2}{dX^2}\left[(1-\alpha X)(1-\beta X)^3\frac{a}{L^2}\left(12 - 24X + 12X^2\right)\right] - \frac{qL^4}{EI_0} \tag{6.88}$$

Performing the differentiation

$$\varepsilon = \frac{12a}{L^4}\begin{bmatrix} 2(1-\alpha X)(1-\beta X)^3 + 12\beta(1-X)(1-\alpha X)\left(1-\beta X^2\right) \\ +4\alpha(1-X)(1-\beta X)^3 + 6\beta^2(1-X)(1-\beta X)\left(1-2X+X^2\right) \\ +6\alpha\beta(1-\beta X)^2\left(1-2X+X^2\right) \end{bmatrix} - \frac{qL^4}{EI_0} \tag{6.89}$$

Orthogonalizing as per Galerkin method, we have

$$\int_0^1 \left\{ \frac{12a}{L^4}\begin{bmatrix} 2(1-\alpha X)(1-\beta X)^3 + 12\beta(1-X)(1-\alpha X)(1-\beta X^2) \\ +4\alpha(1-\alpha X)(1-\beta X)^3 + 6\beta^2(1-\alpha X)(1-\beta X)(1-2X+X^2) \\ +6\alpha\beta(1-\beta X)^2(1-2X+X^2) \end{bmatrix} - \frac{qL^4}{EI_0} \right\} \times \left(6X^2 - 4X^3 + X^4\right)dX = 0 \tag{6.90}$$

There are six integrals in the above, they are evaluated as follows:

$$I_1 = \int_0^1 \left\{ \frac{12}{L^4} \left[2(1-\alpha X)(1-\beta X)^3 \right] \right\} (6X^2 - 4X^3 + X^4)dX$$
$$= \frac{12}{L^4} \left\{ \frac{12}{5} - \frac{26}{15}\alpha - \frac{26}{5}\beta + \frac{142}{35}\alpha\beta + \frac{142}{35}\beta^2 - \frac{93}{28}\alpha\beta^2 - \frac{31}{28}\beta^3 + \frac{59}{63}\alpha\beta^3 \right\} \tag{6.91a}$$

$$I_2 = \int_0^1 \left\{ \frac{12}{L^4} \left[12\beta(1-X)(1-\alpha X)\left(1-\beta X^2\right) \right] \right\} (6X^2 - 4X^3 + X^4)dX$$
$$= \frac{12}{L^4} \left\{ 4\beta - \frac{16}{7}\alpha\beta - \frac{32}{7}\beta^2 + \frac{103}{35}\alpha\beta^2 + \frac{103}{70}\beta^3 - \frac{43}{42}\alpha\beta^3 \right\} \tag{6.91b}$$

$$I_3 = \int_0^1 \left\{ \frac{12}{L^4} \left[4\alpha(1-\alpha X)(1-\beta X)^3 \right] \right\} (6X^2 - 4X^3 + X^4)dX$$
$$= \frac{12}{L^4} \left\{ \frac{4}{3}\alpha - \frac{16}{7}\alpha\beta + \frac{103}{35}\alpha\beta^2 - \frac{43}{126}\alpha\beta^3 \right\} \tag{6.91c}$$

$$I_4 = \int_0^1 \left\{ \frac{12}{L^4} \left[6\beta^2(1-\alpha X)(1-\beta X)\left(1-2X+X^2\right) \right] \right\} (6X^2 - 4X^3 + X^4)dX$$
$$= \frac{12}{L^4} \left\{ \frac{6}{7}\beta^2 - \frac{57}{140}\alpha\beta^2 - \frac{57}{140}\beta^3 + \frac{47}{210}\alpha\beta^3 \right\} \tag{6.91d}$$

$$I_5 = \int_0^1 \left\{ \frac{12}{L^4} \left[6\alpha\beta(1-\beta X)^2\left(1-2X+X^2\right) \right] \right\} (6X^2 - 4X^3 + X^4)dX$$
$$= \frac{12}{L^4} \left\{ \frac{6}{7}\alpha\beta - \frac{57}{70}\alpha\beta^2 + \frac{47}{210}\alpha\beta^3 \right\} \tag{6.91e}$$

$$I_6 = \int_0^1 \frac{qL^4}{EI_0} (6X^2 - 4X^3 + X^4)dX = \frac{6qL^4}{5EI_0} \tag{6.91f}$$

Substituting the above integrals in (6.90) and simplifying, we obtain the constant a

$$a = \frac{\frac{6}{5}qL^4}{12EI_0\left(\frac{12}{5} - \frac{2}{5}\alpha - \frac{6}{5}\beta + \frac{12}{35}\alpha\beta + \frac{12}{35}\beta^2 - \frac{9}{70}\alpha\beta^2 - \frac{3}{70}\beta^3 + \frac{2}{105}\alpha\beta^3\right)} \tag{6.92}$$

We then have the solution for deflection from (6.36)

$$w(X) = \frac{\frac{6}{5}qL^4(6X^2 - 4X^3 + X^4)}{12EI_0\left(\frac{12}{5} - \frac{2}{5}\alpha - \frac{6}{5}\beta + \frac{12}{35}\alpha\beta + \frac{12}{35}\beta^2 - \frac{9}{70}\alpha\beta^2 - \frac{3}{70}\beta^3 + \frac{2}{105}\alpha\beta^3\right)} \tag{6.93}$$

For a uniform beam with $\alpha = \beta = 0$, the above reduces to

$$w(X) = \frac{qL^4}{24EI}\left(6X^2 - 4X^3 + X^4\right) \tag{6.94}$$

This is same as (6.41) obtained for a uniform beam.

It is the computer that has revolutionized the way in which design problems are carried out in industry, be it Solids, Fluids, Electromagnetics, Heat Transfer. Since the commercial codes survive on markets, they are also developed based on the most common needs in design. The finite element methods depend on a solver that most commonly solves a huge size problem of solving homogeneous or nonhomogeneous set of linear equations. By hand calculations we can use Kramer's rule with three simultaneous linear equations. Anything beyond a size of four equations that is a matrix of size 4 $\times$ 4 is going to be difficult if not impossible. That's where the computers score over human brain and capacity in number crunching. The commercial codes are usually developed in three sections:

1. Preprocessing
2. Solver
3. Post processing

1. Preprocessing: As the name implies, here the finite element model is prepared. First we prepare the geometry of the structure and then convert it to a solid model. There are several meshing tools in the Preprocessor; for example, Hyper Mesh is one of the most popular ones that are used for meshing. The mesh could be two noded 1D elements, three noded triangular plate elements, quadrilateral elements, 8 noded solid elements … Such a model is ready for computer aided design and therefore called CAD model or Meshed model. We then assign boundary conditions and loads. The CAD model thus prepared goes to a solver.
2. Solver: Invariably Solvers are rugged well proven matrix equation solvers with very huge size that can run to few million nodes, depending on how intricate and complex is the geometry. One can use any commercially available platforms or the ones that your CAD laboratory provides.
3. Post processor: Post Processor as the name indicates processes the results of three displacements, six strains, six stresses, principal stresses, Von Mises Stress etc. at each node of the finite element model and displays in a suitable way you choose.

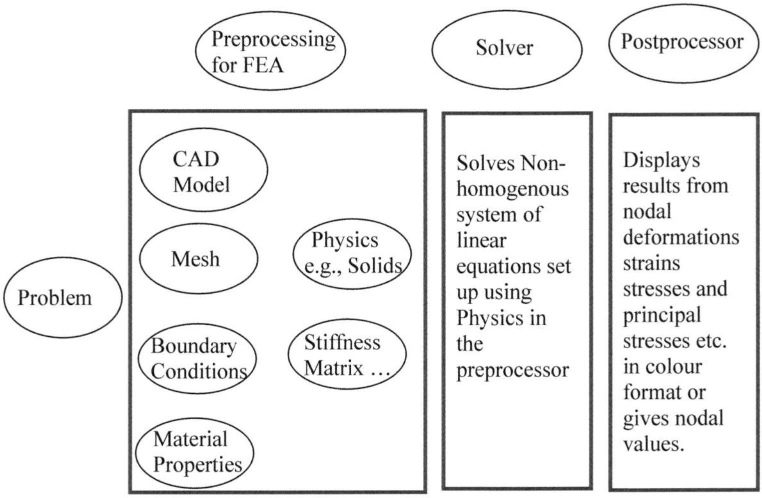

Fig. 6.10 Basics of commercial codes

Figure 6.10 schematically gives the basics of the commercial codes and how they are organized. The main technology aspect is concerned with the elemental stiffness which has been described in this book; the remaining are all concerned with practicing the tool you choose. These tools all come with brochures with on-line help. Just the way one uses Microsoft tools to prepare different documents, the commercial codes are also user friendly with pull down menus. Any student can learn them without attending a course (though a course like this can accelerate). Each student is advised to have fun in solving simple problems that are provided in brochures and then extend them to any problem.

You may begin with one D elements and use the element library to make other complex models. However, you may need to attend advanced strength of materials course dealing with plates, shells and solids to understand how their elemental stiffness matrices are set up. Otherwise, you can just have fun in solving structural problems.

The tree in Fig.6.10 explains the basics of using commercial solvers.

You can use any commercial solver and you will find the above steps in the same.

The commercial codes all have manuals which instruct you how to go through the above steps. The action icons all come out from a pull-down menu and they are quite user friendly. The manuals are similar to any other manual, e.g., that of TV or laptop computer. One can easily learn them by following the instructions. The theory or Physics is something that may not be explained in these manuals and it is for the user to know this part. If one has no idea of Physics of the problem it is likely wrong results and deductions can be made. Remember the procedure is same for a few elements or large number of elements; commercial codes come to rescue when the model size is very large as in the case of industrial applications.

6.8 General Structures by Commercial Solvers

Here we show how the pressure vessel of Example 2.3 can be considered as a three dimensional structure, a pressure vessel shown in Fig. 6.11. The thickness of this pressure vessel can be considered small. So we use what are called *shell elements*. Internal pressure is p = 1.6 MPa, inner diameter is d_i = 5170 mm, cylinder length of pressure vessel is L = 8000 mm and plate thickness is t = 30 mm.

Mean Radius = $\frac{d_i}{2} + \frac{t}{2} = 2.6$ m

$$\sigma_l = \frac{pr}{2t} = 69.33 \text{ MPa} \quad \sigma_h = \frac{pr}{t} = 138.667 \text{ MPa}$$

$$\sigma_{\text{eqv}} = \sqrt{(\sigma_l)^2 + (\sigma_h)^2 - \sigma_l \times \sigma_l} = 120.08 \text{ MPa}$$

The geometric model with quarter-symmetry is shown in Fig. 6.12.

The FEA model is given in Fig. 6.13.

The boundary conditions applied are given in Fig. 6.14.

The Equivalent (Von Mises) Stress obtained is given in Fig. 6.15.

The result obtained in Worked Example 2.3 is 120.08 MPa and it agrees well with the Finite element result 122.94 MPa above. This is an example where the result could be obtained by a simple equilibrium model approach for the required section; however, in complex geometry systems, the commercial solvers with large elements help the designer.

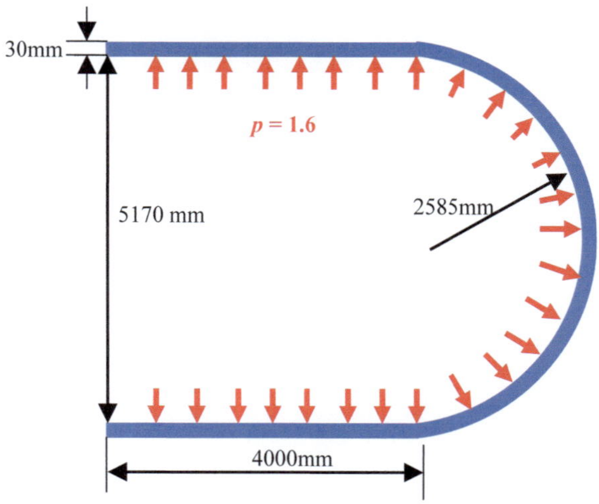

Fig. 6.11 Pressure vessel

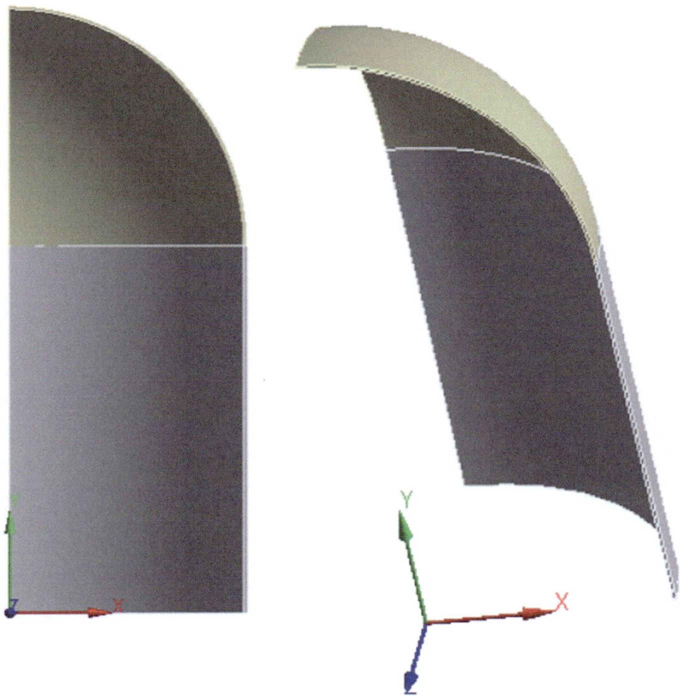

Fig. 6.12 Geometric model

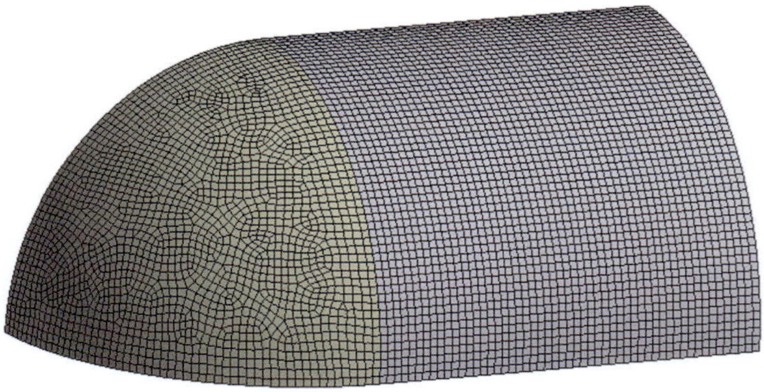

Fig. 6.13 FEA model

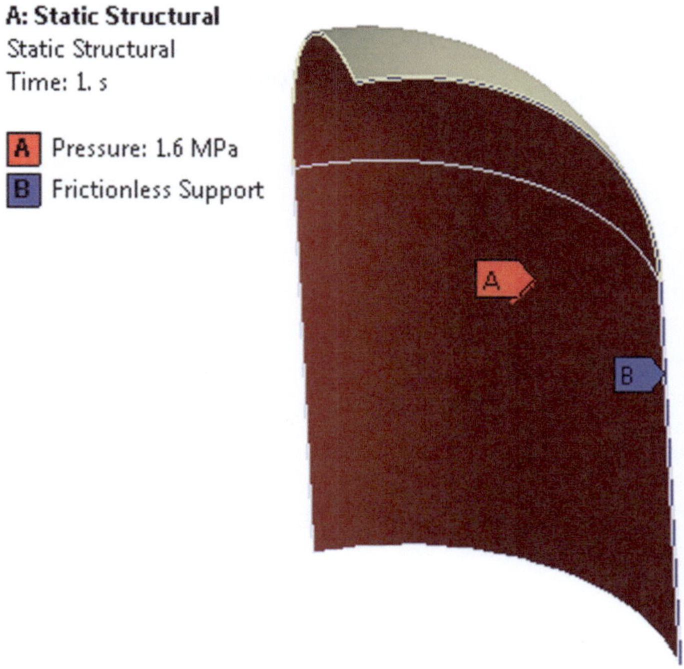

Fig. 6.14 Boundary conditions

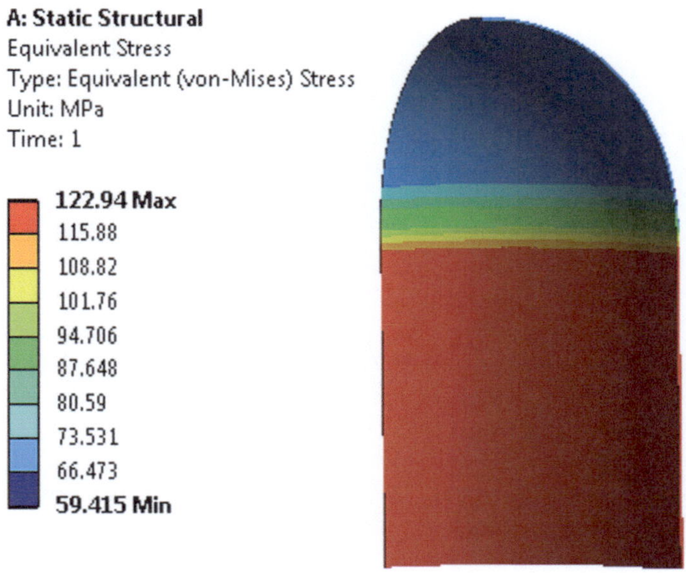

Fig. 6.15 Equivalent stress

6. Exercises

6.1 Define a beam and explain the underlying assumptions in the displacement, strain and stress fields.

6.2 Show that the stress is axial in a bending beam due to transverse loads.

6.3 A 4 m cantilever beam of circular cross-section diameter equal to 10 cm is subjected to a 50,000 N transverse load at the free end. Take E = 200 GPa. Determine maximum stress, its location and direction.

6.4 A 5 m cantilever beam carries a vertical load at the free end. Its cross-section is rectangular with breadth equal to a fifth of the depth. Determine the cross-section dimensions if the maximum deflection is not to exceed 3 mm. Take E = 200 GPa.

6.5 A cantilever beam carries a uniformly distributed load; derive a relation for its deflection at any distance and the maximum deflection. If the uniformly distributed load is 300 N/m, flexural rigidity 60 M Nm^2 determine maximum deflection and slope of the beam.

6.6 A simply supported beam 7.5 m long has a central load 500,000 N. The deflection at the point of load application is 2 mm. What is the slope at the supports?

6.7 A hollow circular cross-section simply supported beam with the outer diameter equal to 80 mm and the inner diameter equal to 50 mm. The beam is 6 m long with a load of 1000 N acting vertically down at its middle. The Young's modulus is 205 GPa. Determine the maximum deflection.

6.8 Show that the maximum deflection of a simply supported beam carrying a uniformly distributed load w N/m is $\frac{5}{384}\frac{wl^4}{EI}$.

6.9 A simply supported steel beam is made from a hollow tube 8 cm outer dia and 4 cm inner dia and has a span of 6 m. The density is 7300 kg/m^3 with E = 205 GPa. What will be deflection at the middle due to its self weight.

6.10 Use Ritz method to find the deflection in problem 6.9 and compare the result obtained and make any comments on the energy method.

6.11 A 6 m long beam with flexural rigidity = 350 M Nm^2 is simply supported and carries a uniformly distributed load 2000 N/m and a concentrated load 25,000 N at 1/3 distance from left support. What is the slope and deflection at the midpoint along the beam?

6.12 A 3 m long cantilever beam carries a uniformly distributed load 500 N/m and in addition a point load 1000 N at 1.6 m distance from the fixed end. If the tip deflection is not to exceed 2 mm what flexural rigidity should be used for the cantilever?

6.13 Using Macaulay's method, determine the deflection equation of a fixed-fixed beam with a midpoint load. If the load is uniformly distributed what would be the equation for deflection and slope along the beam.

6.14 Use energy method for the above problem and compare the results.

6.15 An axially loaded steel bar 6 m long of rectangular cross-section is tapered from 3 cm width and 1 cm thick at fixed end to a point at the free end in a linear manner. The free end is pulled by a force 1000 N. Using energy method determine the axial deflection and stress along the axis of the bar.

Chapter 7
Epilogue

The world of science changed our lives from the 17th century. The science revolution has brought a culture of scientific thinking through parameterization of Physics that we have been observing over millennia. It brought us the realm of engineering directly to our doorsteps. Industrial revolution became a reality. Initially design remained to deal with engineering practice and testing and with the advent of rotating machinery at the turn of 20th century when analytical design became a necessity. Unfortunately we did not have computational capabilities beyond using logarithmic tables and hand calculations.

Therefore the 15 coupled partial differential equations of solids were simplified drastically to one dimension and approximate methods were evolved to get along with the design of rotating machinery and supporting structures by using a correction factor (Factor of Safety) to account for the approximations.

This scenario changed once high performance computing became common. This has brought back fundamental tenets of scientific revolution over three centuries. Yet many educational institutions continue to emphasize the century old approximations rather than the current industrial practices. This has meant that the commercial codes are just tools and very rarely engineers of the day know what happens in a code. The codes are merely replacing log tables or slide rules of yester year, using the same science, more accurately rather than through approximations.

This book brings out an approach which directly brings science to the fore that makes the young mind understand what the commercial codes are doing and bring students to current engineering practices more intelligibly; Simulation Based Engineering Science (SBES) Approach.

J.S. Rao, *Simulation Based Engineering in Solid Mechanics*,
DOI 10 1007/978-3-319-47614-8_7

Index

J.S. Rao, *Simulation Based Engineering in Solid Mechanics*,
DOI 10.1007/978-3-319-47614-8

MIX
Papier aus verantwortungsvollen Quellen
Paper from responsible sources
FSC® C105338

If you have any concerns about our products,
you can contact us on
ProductSafety@springernature.com

In case Publisher is established outside the EU,
the EU authorized representative is:
Springer Nature Customer Service Center GmbH
Europaplatz 3, 69115 Heidelberg, Germany

Printed by Libri Plureos GmbH
in Hamburg, Germany